COMMUNITY POWER

SOME OTHER VOLUMES IN THE SAGE FOCUS EDITIONS

6. **Natural Order**
   Barry Barnes and Steven Shapin
8. **Controversy (Second Edition)**
   Dorothy Nelkin
14. **Churches and Politics in Latin America**
   Daniel H. Levine
21. **The Black Woman**
   La Frances Rodgers-Rose
24. **Dual-Career Couples**
   Fran Pepitone-Rockwell
31. **Black Men**
   Lawrence E. Gary
32. **Major Criminal Justice Systems**
   George F. Cole, Stanislaw J. Frankowski, and Marc G. Gertz
34. **Assessing Marriage**
   Erik E. Filsinger and Robert A. Lewis
36. **Impacts of Racism on White Americans**
   Benjamin P. Bowser and Raymond G. Hunt
41. **Black Families**
   Harriette Pipes McAdoo
43. **Aging and Retirement**
   Neil G. McCluskey and Edgar F. Borgatta
47. **Mexico's Political Economy**
   Jorge I. Dominguez
50. **Cuba**
   Jorge I. Dominguez
51. **Social Control**
   Jack P. Gibbs
52. **Energy and Transport**
   George H. Daniels, Jr., and Mark H. Rose
54. **Job Stress and Burnout**
   Whiton Stewart Paine
56. **Two Paychecks**
   Joan Aldous
57. **Social Structure and Network Analysis**
   Peter V. Marsden and Nan Lin
58. **Socialist States in the World-System**
   Christopher K. Chase-Dunn
59. **Age or Need?**
   Bernice L. Neugarten
60. **The Costs of Evaluation**
   Marvin C. Alkin and Lewis C. Solmon
61. **Aging in Minority Groups**
   R.L. McNeely and John N. Colen
62. **Contending Approaches to World System Analysis**
   William R. Thompson
63. **Organizational Theory and Public Policy**
   Richard H. Hall and Robert E. Quinn
64. **Family Relationships in Later Life**
   Timothy H. Brubaker
65. **Communication and Organizations**
   Linda L. Putnam and Michael E. Pacanowsky
66. **Competence in Communication**
   Robert N. Bostrom
67. **Avoiding Communication**
   John A. Daly and James C. McCroskey
68. **Ethnography in Educational Evaluation**
   David M. Fetterman
69. **Group Decision Making**
   Walter C. Swap and Associates
70. **Children and Microcomputers**
   Milton Chen and William Paisley
71. **The Language of Risk**
   Dorothy Nelkin
72. **Black Children**
   Harriette Pipes McAdoo and John Lewis McAdoo
73. **Industrial Democracy**
   Warner Woodworth, Christopher Meek, and William Foote Whyte
74. **Grandparenthood**
   Vern L. Bengtson and Joan F. Robertson
75. **Organizational Theory and Inquiry**
   Yvonna S. Lincoln
76. **Men in Families**
   Robert A. Lewis and Robert E. Salt
77. **Communication and Group Decision-Making**
   Randy Y. Hirokawa and Marshall Scott Poole
78. **The Organization of Mental Health Services**
   W. Richard Scott and Bruce L. Black
79. **Community Power**
   Robert J. Waste

COMMUNITY POWER

Directions for Future Research

Edited by

Robert J. Waste

SAGE PUBLICATIONS
The Publishers of Professional Social Science
Beverly Hills Newbury Park London New Delhi

For information address:

SAGE Publications, Inc.
275 South Beverly Drive
Beverly Hills, California 90212

SAGE Publications Inc.
2111 West Hillcrest Drive
Newbury Park
California 91320

SAGE Publications Ltd.
28 Banner Street
London EC1Y 8QE
England

SAGE PUBLICATIONS India Pvt. Ltd.
M-32 Market
Greater Kailash I
New Delhi 110 048 India

Printed in the United States of America

Library of Congress Cataloging-in-Publication Data

Main entry under title:

Community power.

(Sage focus editions ; v. 79)
Includes bibliographies.
1. Community power. 2. Pluralism (Social sciences)
3. Elite (Social sciences). 4. Power (Social sciences)
I. Waste, Robert J.
HN49.P6C66 1986 303.3 85-22275
ISBN 0-8039-2572-7
ISBN 0-8039-2573-5 (pbk.)

FIRST PRINTING

To Susan A. MacManus:
"Blessed be the peacemakers . . . "
Matthew 5.09

Contents

Contents

Acknowledgments

This book evolved as a direct result of a theme panel organized by Professor Susan A. MacManus (Cleveland State University) for the 1984 Annual Meeting of the American Political Science Association. The authors are indebted, singly and jointly, to Ms. MacManus for her vision and tact in successfully bringing together a body of scholars who, as Susan noted, "disagree with each other and are not afraid to say so." Although the disagreements among community power researchers are legion, the present volume concentrates—to a considerable extent—on an emerging body of agreement among these same scholars; an agreement about the shape and directions of urban research in the foreseeable future.

I should like to express my own thanks to San Diego State University and the Faculty Grant-In-Aid program for financial assistance. I would also like to thank the people at Sage Publications for encouragement and faith in the initial proposal, and my wife, Kathrine, for her constant support and goodwill. The latter, it should be noted, was no small task considering that the delivery of this book to the publisher and Kathy's delivery of our first child were less than two weeks apart.

—*R.J.W.*

PART I

Introduction

1

Community Power

Old Antagonisms and New Directions

ROBERT J. WASTE

The pluralist-elitist "impasse" (Ricci, 1971) is, as Clarence Stone (1981: 505) has noted, "now entering its third decade." In some respects this quarrel between pluralist and elite analysts of community power is the social science equivalent of the feud between the Hatfields and the McCoys. In fact, the analogy to the Hatfield-McCoy feud is, in three respects, well placed. First, adherents of both camps have exchanged a great deal of fire (Dahl, 1961, 1963, 1966, 1978, 1979, 1982; Domhoff, 1967, 1978; Dye, 1976; Hunter, 1953, 1980; Polsby, 1959, 1960a, 1969, 1972, 1979, 1980; Ricci, 1971, 1984; Stone, 1976; Walker, 1966; Wolfinger, 1960, 1971, 1974). Second, the feud—circa 1960 to the present—has exhibited considerable staying power. Third, the staying power of the feud depends, at least in part, on the successful depiction of the other side's claims in terms of inaccurate or timeworn stereotypes.

The present volume, although not intended to resolve completely the pluralist-elitist community power dispute, does hope to defuse the conflict and redirect it in a more positive direction. Serious and legitimate differences separate adherents of the decisional, reputational, and positional approaches to the study of community power (Ricci, 1971, 1984). Having said this, it is also necessary to add that less legitimate impediments also separate pluralist and elitist students of community power. Foremost among these are oversimplifications and misrepresentations of the positions held by the antagonists themselves. Further fueling the impasse fires are, as we shall argue, the insularity (until very recently) of the main actors in the dispute, and the lack (again, until very recently) of innovation in both community power methodologies and theoretical conceptualization. This volume is both a consequence and a continuation of recent efforts to achieve a certain measure of détente in community power research.

MAJOR ACTORS AND POSITIONS

The major approaches to community power studies may, as David Ricci has suggested, be divided into three camps. These are the "reputational" approach of Hunter (1953), the "positional" approach associated with Mills (1958), and the "decisional" approach developed by Dahl (1961).

The Elite Approach

The reputational and positional approaches are variants of the elite approach to studying community power. The two approaches share similar assumptions about political power in local communities, but each elite variant has developed a unique approach to studying local decision making. Elite analysts—notably Hunter, Mills, Dye, and Domhoff—assume that in virtually all communities a relatively small group of individuals exercises control over dominant resources and personnel, and controls the outcome of all key decisions within the community. For Hunter, the central task of studying power in local communities consisted of identifying the principal "leaders"—the local power brokers. Hunter argued that "the difference between the leaders and other men lies in the fact that social groupings have apparently given definite social functions over to certain persons and not to others" (1953:

2). Hunter developed the reputational method to identify such community leaders. The reputational method consists of (1) listing all persons presumed to exercise influence in local business, government, civic, or social circles; (2) asking a cross-section of community "knowledgeables" to rank those on the list, and delete or add names where appropriate; and, finally, (3) reducing the list to manageable proportions for a community power study. In the case of Atlanta, for example, the list was pared down from 175 to 40 leaders. Thus the heart of the reputational method—a method criticized in some quarters as "nomination by rumor, confirmation by innuendo"—remains "listing persons of repute and then interviewing to validate the list" (Ricci, 1971: 90).

The positional or "power elite" variant developed by Mills and associated with such contemporary community power scholars as Domhoff and Dye shares the assumption that political power in most communities is exercised by a relatively few players. However, rather than seeking leaders by nomination à la Hunter, the positional analysts prefer to identify major political, economic, military, or social associations or institutions, and then to analyze the membership, cohesiveness, and impact of such groups. Of particular importance to the positional analysts are the overlapping memberships that such community leaders, usually but not exclusively wealthy leaders, share with other members of their class. What emerges from this view is a picture of a ruling class of wealthy and highly placed individuals who have a predominant role in both the national and local decision-making process.

In addition to the seminal study by Mills, *The Power Elite* (1956), which first advanced the notion that a ruling class comprising the leaders of key economic, social, and military groups controls the national policy process on all key issues of crucial importance to them as a class, several other major positional studies have been conducted. Although full discussions of the positional studies are available in print (Ricci, 1971, 1984; Polsby, 1980), we should briefly recognize the leading contributions of two of the present chapter contributors, G. William

Domhoff and Thomas R. Dye. Both Domhoff and Dye have made a career of attempting to identify key social groups—notably the wealthy—and tracing their membership and influence. Domhoff has analyzed the role of money and the Democratic Party (1972), the role of social class and social clubs (1970, 1974), and in *Who Rules America?* (1967), has extended and updated the power elite thesis of Mills. Dye has also focused on the impact of wealth and ruling class cohesiveness in such studies as *Who's Running America: The Carter Years* (1976), *Who's Running America: The Reagan Years* (1983), and in studies of the policymaking process at the local level (Dye, 1985; Dye and Renick, 1981).

The Pluralist Approach

The pluralist, or decisional, approach to community power evolved out of a study conducted in 1957-1959 in New Haven by Robert A. Dahl of Yale University and two graduate assistants, Nelson Polsby and Raymond Wolfinger. Dahl's *Who Governs? Democracy and Power in an American City* (1961), Polsby's *Community Power and Political Theory* (1963), and Wolfinger's "Reputation and Reality in the Study of Community Power" (1960) form the core of the pluralist approach to community power. Dahl and his associates rejected "reputation"—or the reputational approach of Hunter—in favor of "reality" in community power studies. Dahl focused on "public governments" (1963: 254) and their expenditures and decisions rather than the positional or reputational approach of Hunter and Mills in which actors in major social, economic, military, and political hierarchies and associations are identified and their influence gauged. Dahl identified several criteria for determining "important" decisions worth studying in a community power analysis. As Polsby describes it, the New Haven decisions studied were selected and ranked using four criteria:

(1) how many people are affected by outcomes,
(2) how many different kinds of community resources are distributed by outcomes,

(3) the amount of resources distributed by outcomes, and
(4) how drastically present community resource distributions are altered by outcomes (Polsby, 1980: 96).

More important, "Dahl assumed that in New Haven, as in any other community, only a few . . . [actors] would actually make important decisions; this reality he conceded to Hunter and Mills" (Ricci, 1971: 131). These actors—drawn from the politically active (the "homo politicus" strata) in New Haven—were formed by the researchers into "leadership pools" of "all persons formally connected with decision making" (Polsby, 1959: 798) in the areas chosen for study. What Dahl disputed with the elitists was not the *number* of people involved in community decisions so much as the *scope* of their influence, and the *interplay* between decisions made by citizens voting in local elections and decisions made by those in the various community leadership pools.

After studying public education, urban renewal, and party nominations in New Haven, Dahl concluded—in a view that, as Ricci noted (1971: 175), "flatly contradicted virtually every significant notion advanced in Hunter's *Community Power Structure* and Mills's *The Power Elite*"—that the leadership pools that affected public policy in New Haven in one arena (public eduction, for example) did *not* overlap into other arenas (such as party nominations or urban renewal). Further, the "influentials" in policy arenas such as education, nominations, and renewal bore little or no resemblance to the elites found by the reputational or positional scholars. Dahl found very little overlap between the policy elites or policy influentials and the social elites—the social and economic "notables" of New Haven. Emphasizing the centrality of this finding, Dahl writes:

> Probably the most striking characteristic of influence in New Haven is the extent to which it is *specialized*; that is, individuals who are influential in one sector of public activity tend not to be influential in another sector; and, what is probably more significant, the social strata from which individuals in one sector tend to

> come are different from the social strata from which individuals in other sectors are drawn [1961: 169].

Finally, Dahl's description of New Haven included a clear indication that electoral politics was important in New Haven, and that both voters and decision makers operated with elections in mind.

Dahl's pluralist New Haven, then, stood in stark contrast to Hunter's elitist Atlanta. With little apparent room for compromise, an impasse soon developed between adherents of the two camps (Ricci, 1971: 205-216).

IMPASSE, INSULARITY, AND STEREOTYPES

By 1967, this impasse had become strident. Several major studies demonstrated that the lines separating the two camps were now drawn sharply, and with minimal opportunity for rapprochement. Prominent in this period were an avalanche of studies criticizing the pluralist approach to studying community power (D'Antonio and Ehrlich, 1961; Kariel, 1961, 1970; Anton, 1963; Bachrach and Baratz, 1963; Connolly, 1967; Domhoff, 1967; McCoy and Playford, 1967). The issue, in turn, was joined by the pluralists who responded in quantity and in kind (Wolfinger, 1960; Dahl, 1963, 1966; Polsby, 1963, 1969, 1980). One indication of the tenor of the debate during this period may be seen in an exchange between Polsby and Domhoff. Polsby, in a review of *Who Rules America?* (1968), a positional study of the national policy process, labeled Domhoff's work "amateur sociology," yielding "no light whatsoever on how anything works" (Polsby, 1968: 477). In *Bohemian Grove and Other Retreats* (1974: 110), Domhoff reversed and returned the assessment with something of a backhanded compliment when he observed that "the pluralists' single-minded way of studying power has a long and honorable history in American social science." Given such an exchange, it is not difficult to imagine how the pluralist-elitist dialogue became the pluralist-elitist impasse.

The pluralist-elitist standoff, circa 1961-1981, produced and perpetuated a period of insularity in community power studies. Both sides communicated very little with scholars of the opposing side, and, second, the focus of community power studies became somewhat insular itself—locked into a defense or attack of one's own or one's opponents' methodology. And as in any war, the opponents and their claims were sometimes reduced to stereotypes or strawmen, thus adding animus, if not logic, to the conflict.

The two sides were not unlike two warring navies, each capable of blockading the other's ports, but neither side possessed of sufficient firepower to end the conflict. Locked inside these ports—New Haven and Atlanta, respectively—community power studies engaged in a lengthy (1961-1985) dispute. Much of this debate was, in retrospect, insular in nature. It turned either on the assertion that Atlanta ("Regional City") was elitist (Hunter, 1953, 1980; Stone, 1976; Domhoff, 1978) or that New Haven was characterized by pluralist politics (Dahl, 1961; Polsby, 1980; Wolfinger, 1960).

Pluralist or elitist scholars of this period occasionally shifted focus to locales other than New Haven or Atlanta. Notable among these studies are Wildavsky (1964), Crenson (1971), Eulau and Prewitt (1973), Wirt (1974), and Gaventa (1980). These studies avoided the insularity of focusing on the heavily studied New Haven-Atlanta sites and contained theoretical developments foreshadowing more recent changes in community power studies. Especially helpful was the distinction by Eulau and Prewitt between the politics of urban amenities and economic growth in the San Francisco Bay Area, the analysis of "hyperpluralism" in San Francisco city politics by Wirt, and the study of the politics of powerlessness in Appalachian communities by Gaventa. Due in part to these and other studies of the period, the insular and strawman quality of community power studies began to thaw in the mid-1970s.

THE "COLD WAR" THAWS

Somewhere around the mid- to late 1970s, détente in community power research (as this book proves) became possible. Several factors seem significant in this development. Four that we shall treat here are: (1) "a powerful resurgence in the application of quasi-Marxian insights . . . to all manner of American settings, not the least the local community" (Polsby, 1980: xi); (2) new theoretical developments in conceptualizing important questions in community power and urban studies; (3) new advances in the methodology of studying community power; and (4) a general climate among community power researchers conducive to a limited form of rapprochement. Let us consider each of these developments in turn.

First, as Polsby notes in the Preface to the second edition of *Community Power and Political Theory* (1980), there is indeed a renewed interest in both community power and Marxist approaches to the study of power in local communities. As Domhoff notes in Chapter 3, the critique of Marxist assumptions about the distribution of power in local communities—most notably by Molotch (1976, 1979, 1984)—has served as a catalyst for renewed interest in examining community power. An additional and important indicator of a renaissance of interest in community power is the increasing number of recent community power studies and articles (Polsby, 1980; Falkemark, 1982; Domhoff, 1983; Manley, 1983; Bolland, 1984; Ghiloni, 1984; Ghiloni and Domhoff, 1984; Kurtz, 1984; Mizrachi, 1984; Stone, 1984; Wirt, 1984; Waste, 1986).

Second, several recent innovative attempts to rethink basic theoretical problems in community power and urban studies have spawned a great deal of interest. These include Stone's attempt to combine decisional and positional methodologies in studying redevelopment in Atlanta (1976), and his more recent discussion of "systemic power" in local communities (1980). Polsby has contributed to this rethinking of the theoretical underpinnings of community power studies with a careful reexamination of the thorny issues of "non-decision-making" and "objective interests" in local communities (1980). Dahl (1982: 31-54) has challenged traditional thinking on the problems of pluralist democracy. Wirt, in his classic study of hyperpluralism in San Francisco (1974), and more recently in "Rethinking Community Power" (1984), has served to sharpen the theoretical focus of the study of local governance. A final and important theoretical contribution, as Dye underlines in Chapter 2, is the development by Peterson (1981) of a typology of local policymaking that can serve as a productive benchmark for both pluralist and elite community power scholars.

A third impetus to renewed interest in community power studies are several recent attempts to rethink the field methodologies appropriate for positional, reputational, or decisional researchers. Recent methodological studies by reputational (Hunter, 1980; Trounstine and Christensen, 1982), positional (Falkemark, 1982; Domhoff, 1983), and decisional scholars (Polsby, 1980; Waste, 1986) indicate a high level of interest in refining the ways in which community power researchers currently study the local government laboratory.

Finally, a general climate of peace was attained among the community power researchers (for example, Polsby, 1980: 236-237; Domhoff, 1983: 203-223) by the development of a short-lived journal, *Power and Elites* by Dye and Domhoff, that genuinely attempted to balance elitist and pluralist accounts of community power. The formation of a theme panel for the 1984 annual meeting of the American Political Science Association, in which five of the present chapter contributors—Dye, Domhoff, Eulau, Stone, and Waste—participated, also seems to suggest a renewed vigor and interest in community power studies.

CONCLUSION

The 1975-1981 period in community power research has seen the emergence of a fragile détente in community power studies. This détente has made possible the present volume in which leading elite and pluralist scholars each present original essays refining important questions of theory and method, as well as attempting to forge links between the study of community power and the broader study of local government and general public policy. Representative of this new effort to link community power studies and—more broadly conceived—policy studies is Chapter 2, by leading policy scholar Thomas R. Dye. Chapter 3, by the positional scholar G. William Domhoff, sketches out new theoretical dimensions for elite analysts by elaborating a "growth machine" theory of local decision making. Chapters 4 and 5, by Clarence Stone and Robert Waste, respectively, attempt to clarify central terms—for instance, "power" and "pluralism"—in the community power debate. Chapter 6, by Heinz Eulau, demonstrates in a micro study the usefulness of applying network analysis to community power studies. Chapter 7, an interview with Robert A. Dahl, provided America's leading pluralist thinker with an opportunity to reflect on the original New Haven study as well as review more than thirty years of research, and to speculate on profitable directions for future research in community power and the study of local government generally. Finally, in a brief concluding essay, Chapter 8 knits the earlier essays together into a challenging agenda for the future of urban studies and community power research.

REFERENCES

ANTON, T. J. (1963) "Power, pluralism and local politics." Administrative Science Quarterly 7 (March): 425-457.

BACHRACH, P. and M. S. BARATZ (1963) "Decisions and nondecisions: an analytical framework." American Political Science 57 (September): 632-642.

BOLLAND, J. M. (1984) "The limits of pluralism." Power and Elites 1 (Fall): 69-88.

CONNOLLY, E. [ed.] (1969) The Bias of Pluralism. New York: Lieber-Atherton.

CRENSON, M. (1971) The Un-Politics of Air Pollution. Baltimore: Johns Hopkins University Press.

DAHL, R. A. (1961) Who Governs? New Haven: Yale University Press.

———(1963) "Reply to Thomas Anton's 'Power, Pluralism and Local Politics.' " Administrative Science Quarterly 7 (March): 250-256.

———(1966) "Further reflections on 'The Elitist Theory of Democracy.' " American Political Science Review 60 (June): 296-305.

———(1977) "On removing certain impediments to democracy." Political Science Quarterly 92 (Spring).

———(1978) "Pluralism revisited." Comparative Politics 10 (January): 191-204.

———(1979) "Who really rules?" Social Science Quarterly 60 (June): 144-151.

———(1982) Dilemmas of Pluralist Democracy: Autonomy vs. Control. New Haven: Yale University Press.

——— and C. E. LINDBLOM (1953) Politics, Economics, and Welfare. New York: Harper & Row.

D'ANTONIO, W. V. and H. J. EHRLICH [eds.] (1961) Power and Democracy in America. South Bend: Notre Dame University.

DOMHOFF, G. W. (1967) Who Rules America? Englewood Cliffs, NJ: Prentice-Hall.

———(1970) The Higher Circles. New York: Vintage.

———(1972) Fatcats and Democrats. Englewood Cliffs, NJ: Prentice-Hall.

———(1974) The Bohemian Grove and Other Retreats: A Study in Ruling Class Cohesiveness. New York: Harper & Row.

———(1978) Who Really Rules? Santa Monica, CA: Goodyear.

———(1983) Who Rules America Now? A View for the 80's. Englewood Cliffs, NJ: Prentice-Hall.

DYE, T. R. (1976) Who's Running America? The Carter Years. Englewood Cliffs, NJ: Prentice-Hall.

———(1983) Who's Running America: The Reagan Years. Englewood Cliffs, NJ: Prentice-Hall.

———(1984) Understanding Public Policy. Englewood Cliffs, NJ: Prentice-Hall.

———(1985) Politics in States and Communities. Englewood Cliffs, NJ: Prentice-Hall.

———and J. RENICK (1981) "Political power and city jobs." Social Science Quarterly 62 (September): 475-486.

EULAU, H. and K. PREWITT (1973) The Labyrinths of Democracy. Indianapolis: Bobbs-Merrill.

FALKEMARK, G. (1982) Power, Theory and Value. Lund, Sweden: Liber Gleerup.

GAVENTA, J. (1980) Power and Powerlessness. Urbana: University of Illinois Press.

GHILONI, B. W. (1984) "Women, power and the corporation." Power and Elites 1 (Fall): 37-50.

———and G. W. DOMHOFF (1984) "Power structure research and the promise of democracy." Revue Français D'Etudes Americaines 21/22 (November): 335-341.

HUNTER, F. (1953) Community Power Structure. Chapel Hill: University of North Carolina Press.

———(1980) Community Power Succession: Atlanta's Policy Makers Revisited. Chapel Hill: University of North Carolina Press.

KARIEL, H. S. (1961) The Decline of American Pluralism. Stanford: Stanford University Press.

———[ed.] (1970) Frontiers of Democratic Theory. New York: Random House.

KURTZ, D. M. (1984) "Power in Louisiana." Power and Elites 1 (Fall): 51-68.

MANLEY, J. F. (1983) "Neo-pluralism: a class analysis of pluralism I and pluralism II." American Political Science 77 (June): 368-383.

McCOY, C. A. and J. PLAYFORD [eds.] (1967) Apolitical Politics: A Critique of Behaviorism. New York: Crowell.
MILLS, C. W. (1956) The Power Elite. New York: Oxford University Press.
———(1958) "The structure of power in American society." British Journal of Sociology 9 (March).
MIZRUCHI, M. S. (1984) "An interorganizational model of class cohesion." Power and Elites 1 (Fall): 23-36.
MOLOTCH, H. (1976) "The city as a growth machine." American Journal of Sociology 82 (September): 309-330.
———(1979) "Capital and neighborhood in the United States." Urban Affairs Quarterly 14 (March): 289-312.
———(1984) "Romantic Marxism: love is still not enough." Contemporary Sociology 13 (March): 141-143.
PETERSON, P. (1981) City Limits. Chicago: University of Chicago Press.
POLSBY, N. W. (1959) "Three problems in the analysis of community power." American Sociological Review (December): 796-803.
———(1960a) "How to study community power: the pluralist alternative." Journal of Politics 22 (August): 474-484.
———(1960b) "Power in Middletown: fact and value in community research." Canadian Journal of Economics and Social Science 26 (November): 592-603.
———(1968) "Review of 'Who Rules America?'" American Sociological Review (June): 477-478.
———(1969) "'Pluralism' in the study of community power: or erklärung before verklärung in wissenssoziologie." The American Sociologist 4 (May): 118-122.
———(1972) "Community power meets air pollution." Contemporary Sociology 1 (March): 88-91.
———(1979) "Empirical investigation of the mobilization of bias in community power research." Political Studies 27 (December): 527-541.
———(1980) Community Power and Political Theory (2nd ed.). New Haven: Yale University Press. (originally published 1963)
RICCI, D. M. (1971) Community Power and Democratic Theory. New York: Random House.
———(1984) The Tragedy of Political Science: Politics, Scholarship, and Democracy. New Haven: Yale University Press.
STONE, C. N. (1976) Economic Growth and Neighborhood Discontent. Chapel Hill: University of North Carolina Press.
———(1980) "Systemic power in community decision making." American Political Science 74 (December): 978-990.
———(1981) "Community power structure—a further look." Urban Affairs Quarterly 16 (June): 505-515.
———(1984) "The new class or the old convergence?" Power and Elites 1 (Fall): 1-22.
TROUNSTINE, P. J. and T. CHRISTENSEN (1982) Movers and Shakers: A Study of Community Power. New York: St. Martins.
WALKER, J. (1966) "A critique of the elitist theory of democracy." American Political Science 60: 285-295.
WASTE, R. J. (1986) Power and Pluralism in American Cities: Researching the Urban Laboratory. Westport, CT: Greenwood Press.

WILDAVSKY, A. (1964) Leadership in a Small Town. Totowa, NJ: Bedminister Press.
WIRT, F. M. (1974) Power in the City: Decision Making in San Francisco. Berkeley: University of California Press.
———(1984) "Rethinking community power." Power and Elites 1 (Fall): 89-98.
WOLFINGER, R. E. (1960) "Reputation and reality in the study of community power." American Sociological Review 25 (October): 634-644.
———(1971) "Nondecisions and the study of local politics." American Political Science 65 (December): 1063-1104.
———(1974) The Politics of Progress. Englewood Cliffs, NJ: Prentice-Hall.
———and S. ROSENSTONE (1980) Who Votes? New Haven: Yale University Press.

PART II

The Elite View of Community Power

2

Community Power and Public Policy

THOMAS R. DYE

Communities, like nations, are governed by tiny minorities. The community itself, and the lives of the people who live in it, are shaped by a small number of people. "Government is always government by the few, whether in the name of the few, the one or the many" (Lasswell and Lerner, 1952: 7). Elite theory helps us to identify what is truly significant about community life. It helps us to understand communities by describing, clarifying, and simplifying politics, by suggesting explanations of political events and public policies, and by directing inquiry and research in community studies.

The questions in community power research are the universal questions about power in society: Who are the few who have power? How did they acquire their power? How much power do they wield? What are the limits to their power? Do they divide power among themselves, with separate groups making decisions in separate areas, or do they converge into a single hierarchy exercising power in all segments of community life? How open and accessible are these power-holders? Do they compete among themselves over the future of the community, or do they generally share a consensual view of community life? How responsive are these power-holders to the sentiments of the masses of community residents? Can these residents directly influence the decisions of community elites through political organization or electoral participation, or street demonstration? Or are most residents passive, apathetic, and ill-informed? When confronted with adverse community decisions, are residents forced to resign themselves or to "vote with their feet" by moving away, or can residents become politically active and change these decisions?

LIMITS TO COMMUNITY POWER

We know that power is not an attribute of individuals but rather of social institutions. Power adheres to institutional roles which give the people who occupy these roles control over valued resources. At the national level the study of elites involves the identification of positions of authority in large national institutions—institutions that control the economic, governmental, education, civic, and cultural resources of the nation. Most of the nation's economic resources, or "factors of production," are under the control of a small number of large institutions—industrial corporations, banks, utilities, insurance companies, investment firms, and the national government. It makes sense, then, to study the power of these national institutions and the people who occupy positions of authority in them (Dye, 1976; Silk and Silk, 1980; Domhoff, 1983).

But communities are not nations. Community power structures are not like national power structures. Most of the forces shaping life in American communities arise outside of these communities. The ability of any community or its leadership to shape its own destiny is very limited. Community elites cannot make war or peace, or cause inflation or recession, or determine interest rates or the money supply.

Power is centralized in American society in both corporate and governmental institutions. There are 80,000 separate governments in the American political system, but one of these—the U.S. government—collects 68 percent of all governmental revenue. The 50 state governments collect 20 percent of all revenue, and all of the nation's local governments combined collect only 12 percent of all governmental revenues (Tax Foundation, 1984).

Communities are limited in their resources. "City politics is limited politics" (Peterson, 1981: 4).

COMMUNITY POWER STRUCTURES

But there is one "factor of production"—land—which *is* controlled by local elites. Communities are spatially defined social structures. Residents come and go, but the community remains a particular geographic location. Land is a valuable resource; it is a necessary factor of production. Capital must be placed somewhere. Labor and management must be located somewhere. Production requires a spatial location.

Community power structures are composed primarily of landed interests whose goals are to intensify the use of their land and add to the value of it. Community power structures are not miniature versions of national power centers. Community elites seek to maximize rents rather than profits. Rents are broadly defined to include appreciation of land values, real estate commissions, builders' profits, and mortgage interest, increased revenues to commercial enterprises serving the community, as well as rent payments. Community power structures are dominated by mortgage lending banks, real estate developers, builders, and landowners. They may be joined by owners or managers of local utilities, department stores, attorneys and title companies, and others whose wealth is affected by land use. Local bankers who finance the real estate developers and builders are probably at the center of the elite structure in most communities.

Unquestionably these community elites compete among themselves for wealth, profit, power, and preeminence. But they share a consensus about intensifying the use of land. Corporate plants and offices, federal and state office buildings, universities and colleges, all contribute to the increased land values, not only on the parcels used by these facilities but also on neighboring parcels.

Growth is the shared elite value. The community elite is indeed a "growth machine" (Molotch, 1976; also cited by Domhoff, this volume). Economic growth expands the work force and disposable income within the community. It stimulates housing development, retail stores, and other commercial activity. The landed elite understands that they all benefit, albeit to varying degrees, when economic growth occurs within the community.

THE FUNCTION OF COMMUNITY ELITES

The economic function of community elites is to prepare land for capital. Capital investment in the community will raise land values, expand the labor force, generate demand for housing as well as commercial services, and enhance the local tax base. The preparation of land for capital investment involves much more than just providing large tracts of level acreage. It involves the provision of good transportation facilities—highways, streets, rail access, and water and airport facilities. It involves the provision of utilities—water, gas and electrical power, solid waste disposal, and sewage treatment. It involves the provision of good municipal services, especially fire and police protection. It involves the elimination of harassing business regulations and the reduction of taxes on new investments to the lowest feasible levels. It involves the provision of a capable and cooperative labor force, educated for the needs of productive capital and motivated to work. Finally, it includes the provision of sufficient amenities—cultural, recreational, aesthetic—to provide the corporate managers with a desirable life style.

CONSTRAINTS ON POWER

Currently elites are limited in their power to manipulate the factors that affect locational decisions for capital investment. Historically, the most important factor affecting industrial location decisions was access to water transportation. Later, a community's geographical location relative to rail and air transportation and regional markets became a major factor in industrial location. Climate affects the desirability of location, as does terrain. As highway and air transportation began to supplant water and rail transportation, the "Sunbelt" became a more desirable location. Clearly, these geographical factors are not under the control of local elites.

Moreover, elites in different communities must compete with each other to attract capital investment. This competition is a constraint on the power of any particular community elite. Not only must a community compete for new investments, but it must also endeavor to prevent relocation of investments they already have.

Elite Activities

Although constrained by geography, climate, and competition, community elites are not powerless in the pursuit of capital investment. They can minimize local taxes on capital and profits from capital. They can offer public land free of charge or at reduced prices. They can even provide buildings, constructed at public expense, at modest rental prices. They can provide roads, sewers, water, police and fire protection, at or below cost. They can provide good schools for children of both management and labor, and they can provide specialized vocational training in public vocational-technical schools and community colleges to directly assist investing industries. They can ignore spillover costs in air and water pollution or degradation of the environment in order to lower the costs of production. In brief, they can provide a "favorable business climate" for capital investment.

To mobilize mass opinion on behalf of these goals, community elites typically promise jobs as a result of new investment. Growth creates jobs, not only in the basic industries which are the object of elite solicitations to the community, but also in housing construction and commercial and service industries which grow with the community. According to Molotch (1976: 320), jobs are "the key ideological prop for the growth machine."

Elite Consensus

Community elites strive for consensus. They believe that community economic growth—increased capital investment, more jobs, and improved business conditions—benefits the entire community. According to Paul E. Peterson, community residents share a common interest in the economic well-being of the city:

> Policies and programs can be said to be in the interest of cities whenever the policies maintain or enhance the economic position, social prestige, or political power of the city as a whole [Peterson, 1981: 20].

Community elites themselves would doubtlessly agree with Peterson. He adds that the interests of the city as a whole are closely bound to its export industries. These industries add net wealth to the community at large, while support and service industries merely transfer wealth within the community.

> Whatever helps them prosper rebounds to the benefit of the community as a whole—perhaps four or five times over. It is just such an economic analysis (of the multiplier effect of expert industries) that has influenced many local government policies. Especially the smaller towns and cities may provide free land, tax concessions, and favorable utility rates to incoming industries [Peterson, 1981: 23].

The less economically developed a community, the more persuasive the argument on behalf of export industries.

Local government officials are expected to share in the elite consensus. Economic prosperity is necessary for protecting the fiscal base of local government. Growth in local budgets and public employment, as well as governmental services, depends upon growth in the local economy. Governmental growth expands the power, prestige, and status of government officials. Moreover, economic growth is usually good politics. Growth-oriented candidates for public office usually have larger campaign treasuries than antigrowth candidates. Growth-oriented candidates can solicit contributions from the community power structure. Finally, according to Peterson, most local politicians have "a sense of community responsibility." They know that if the economy of the community declines, "local business will suffer, workers will lose employment opportunities, cultural life will decline, and city land values will fall" (Peterson, 1981: 29).

The community power literature confirms that a great many communities are consensual, especially with regard to policies affecting the local economy. Consensus is more likely in smaller communities with a single industry, and independent communities which are not part of a large metropolitan area (Miller, 1958; Miller and Farm, 1960).

Community Conflict and Counterelites

But consensus on behalf of economic growth is sometimes challenged by entrenched community interests. However much the "growth machine" elite may strive for consensus, and despite the admonitions of scholars that economic growth benefits the whole community, some people do not like growth.

Indeed, it has become fashionable in upper-middle-class circles today to complain loudly about the problems created by growth—congestion, pollution, noise, unsightly development, the replacement of green spaces with concrete slabs. People who already own their houses and do not intend to sell them, people whose jobs are secure in government bureaucracies or tenured professorships, people who may be displaced from their homes and neighborhoods by new facilities, people who see no direct benefit to themselves from growth, and business or industries who fear the new competition which growth may bring to the community all combine to form a potentially powerful counterelite.

No-growth movements (or, to use the current euphemism, "growth management" movements) are not mass movements. They do not express the aspirations of workers for jobs or renters for their own homes. Instead, they reflect upper-middle-class aesthetic preferences—the preferences of educated, affluent, articulate homeowners. Growth brings ugly factories, cheap commercial outlets, hamburger stands, fried chicken franchises, and "undesirable" residents. Even if new business or industry would help hold down local taxes, these affluent citizens would still oppose it. They would rather pay the higher taxes associated with no growth than change the appearance or life-style of their community. They have secure jobs themselves and own their homes; they are relatively unconcerned about creating jobs or building homes for less affluent citizens.

No-growth movements dominate many upper-middle-class suburban communities. These communities can enjoy the benefits of economic growth in the metropolitan area at large while protecting themselves from the costs of this growth. They can shift the costs of growth to other communities within the metropolitan area—usually less affluent communities, close-in suburbs, or the central city. What appears to be a "growth management" issue may become in reality a distribution issue within a metropolitan area.

No-growth movements also challenge traditional economic elites in many large and growing cities in the Sunbelt. The no-growth leaders may themselves have been beneficiaries of community growth only five or ten years ago, but they quickly perceive their own interest in slowing or halting additional growth. Now that they have climbed the ladder to their own success, they are prepared to knock the ladder down to preserve their own style of living.

Municipal government offers the tools to challenge the growth elite. Communities may restrict growth through zoning laws, subdivision control restrictions, utility regulations, building permits, and environmental regulations. Opposition to street widening, road building, or tree cutting can slow or halt development. Public utilities needed for development—water lines, sewage disposal facilities, fire houses, and so on—can be postponed indefinitely. High development fees, "impact fees," utility hookup charges, or building permit fees can all be used to discourage growth. Environmental laws, or even "historic preservation" laws, can be employed aggressively to halt development.

Not all of the opposition to growth is upper-middle-class in character. Students of community power have described the struggle of blacks and low-income neighborhood groups in opposing urban renewal and downtown city development (Stone, 1976). However, this literature suggests that traditional community elites are likely to be successful against this kind of opposition. We might speculate that the "growth machine" elites are more concerned about opposition from educated, affluent, upper-middle-class, "growth management" homeowners than they are about opposition from minority, low-income neighborhood groups.

COMMUNITY POLICIES

Community power structures concern themselves primarily with economic growth. But there are many other issues in community politics, some of which hold little interest for landed elites yet generate considerable community controversy. How can we differentiate between issues which arouse the active interest of propertied elites and those which do not?

The literature on community politics provides a starting place for a typology of community issues. Williams and Adrian (1963) distinguished between communities which were concerned with the promotion of economic growth and those which were concerned about preserving life's amenities. Eulau and Prewitt (1973) also developed a distinction between types of community policies: those that promote economic growth ("planning, zoning, urban renewal, attract business, etc.") and those that provide urban amenities ("library, civic center, recreation, etc."). It is interesting that both of these studies focus on the distinction between growth policies and the provision of amenities. We have already suggested that these issues provide the impetus to the formation of the community power structure (growth policies) and of potential counterelites (amenities policies). Peterson (1981: 41) argues convincingly that a typology of community policies helps to identify a policy's "impact on the economic vitality of the community" and to predict the interests of the community's economic elite.

Developmental policies are those which directly enhance the economic position of the community. Developmental policies can be expected to yield economic benefits directly to landed interests in the community and indirectly to many others in the form of more jobs, increased business, and higher tax revenues. Policies to attract industry, build streets and highways, improve transportation facilities, provide utilities, and renew depressed areas are characteristic of developmental policies. Efforts to measure empirically developmental policies of local governments might include expenditures for highways and transportation and urban renewal. Developmental policies may be undertaken by private organizations (chambers of commerce), independent groups of bankers and developers, quasi-governmental bodies (industrial development authorities), as well as by governments themselves.

Redistributional policies are not designed to add to the net economic well-being of the whole community but rather to benefit low-income residents by redirecting the economic resources of the community. Most redistributional policies in the United States—social security, welfare, medical care, unemployment compensation, and so on—are undertaken by state and national governments. Local governments are only occasionally called upon to decide redistributional questions. Many cities provide low-income housing, indigent hospital care, shelters for the homeless, and other services which are only partially paid for by state or federal funds. Frequently, issues arise over the allocation of services to different neighborhoods within the community. Occasionally riots and protest occur and community political systems must respond.

Allocational policies describe the broad range of traditional public services provided by local government. Schools, streets, sewers, water and utilities, garbage collection, parks and recreation are widely allocated throughout the community. There are, of course, frequent distributional arguments over the provision of these services. In general, however, they are provided on a community-wide proportional basis. They are financed largely through property taxes which are proportional in their rates. Occasionally the quality of these services becomes a developmental issue when potential investors express an interest in these services, but, overall, these allocational policies are developmentally and distributionally neutral. At the same time, local government officials testify that much of their time is taken up with allocational concerns (Dye, 1985: 300-523). Most local government expenditures are directed to the provision of these services.

Organizational policies encompass decisions about the structure of decision making in the community. These policies include the composition, terms, and legal activity of school boards, city and county commissions, city managers, school superintendents, independent authorities and boards, and other elected officials. Typical issues may include whether or not to have a city manager, a mayor-council, or a commission form of government, at-large or district elections, partisan or nonpartisan elections, an elected or appointed school superintendent, and so forth. Personnel and employment policies may also be considered organizational—that is, who gets elected to community office and who gets hired by the city. Sometimes these questions become redistributional when issues of minority or low-income representation are raised. Sometimes these questions are intertwined with allocational issues, as when incumbents are charged with poor performance on the job. In contrast, organizational policies deserve a separate category because they consume a great deal of community time and energy and are analytically distinct from the other categories.

COMMUNITY POWER AND TYPES OF POLICIES

The value of a typology of community policies is to help us theorize about the role of community power structures in separate policy arenas. We can now combine our theory about the economic function of community elites with our typology of community policies, and we can speculate about the role of local power structures in each of these policy domains.

Developmental Policy

Community power structures are most directly concerned with developmental policies. So also are the upper-middle-class counter-elites who seek to "manage" growth on behalf of their own aesthetic and life-style values. When community power researchers employ decisional case studies, they usually select developmental issues—urban renewal (Hunter, 1953), civic centers (Agger, Goldrich, and Swanson, 1964), or downtown development (Stone, 1976). Indeed, community power research is largely the politics of community development.

The much maligned "reputational method" of studying community power—asking knowledgeable informants to identify powerful people in the community—is well designed for developmental politics. Hunter (1953: 52) asked a simple question:

> Suppose a major project were before the community that required decisions by a group of leaders that nearly everyone would accept. Which people would you choose, regardless of whether or not you know them personally?

The pluralists attacked this question for failing to specify the "scope" of power:

> Most of the reputational researchers, by their failure to specify scopes in soliciting reputations for influence assume that the power of their leader-nominees is equal for all issues. . . . This is an exceedingly dubious assumption. It is improbable, for instance, that the same people who decide which houses of prostitution are to be protected . . . also plan the public school curriculum [Wolfinger, 1960: 636].

Improbable indeed! This criticism emphasizes the importance of a typology of community policies. But the question itself, and other efforts to solicit *general* attributions of power, appear to direct the respondents' attention to the most important issues confronting the community as a whole, and these are developmental issues. Peterson comments on Hunter's question and on the reputational method in general:

> Notice that this question directs the informants to think about programs and policies that are beneficial to the city as a whole. . . . When the question is phrased this way, the respondent is invited to think about the politics of development [Peterson, 1981: 140].

Reputations for power correlate with the "reality" of power when the issues specified are developmental issues. Peterson (1981: chap. 7) argues that Wildavsky's data on Oberlin (1964) reveal a high correlation between a reputation for power and participation in urban renewal decision making. "Given specific wording of questions used by reputationalists and given the kinds of issues in which the 'power elite' tend to be involved, it seems reasonable to infer that these studies are largely analyses of the politics of development" (Peterson, 1981: 141).

Reputations for power are made in developmental politics. This should not really surprise us. Job, home, and business are viewed by most people as the most important aspects of their lives. Those people in the community who are in a position to influence job opportunities, home building, and industrial and commercial development are viewed by most residents as "powerful." So when residents are asked to make general judgments about power in their community, they name bankers, developers, builders, and businessmen. They do not usually name school board members, civic association officers, neighborhood activists, police or fire chiefs, minority group leaders, or even city councilmembers. These persons may be active in other areas of policymaking (allocational, redistributional, or organizational), but most citizens understand that the really important developmental policies—policies which directly affect jobs, houses, and businesses—will be largely determined by the community power structure.

The community power structure—the mortgage lending banks, real estate developers and builders, and landowners—are directly involved in developmental policies. When they are challenged at all, the challenge originates from upper-middle-class homeowners. Only rarely do lower-income or minority group challenges succeed in modifying developmental policies.

Redistributional Policy

Redistribution policies seldom involve the community power structure directly. Although the resolution of redistribution questions might adversely affect community elites, the policies do not directly confront these questions for several reasons.

First of all, the American federal system largely removes these questions from the agenda of community politics. The major issues in redistribution politics in the United States—social security, medical care for the poor and aged, welfare cash and in-kind benefits, unemployment compensation, minimum wage, and job training—are decided at the federal level. Moreover, local communities could not effectively address these questions even if they wished to do so. For example, citywide minimum wage would drive business outside of the city's boundaries. Otherwise, suburban competitors would confront city business with lower production costs. Attractive welfare services provided by a city would induce the poor from other jurisdictions to come into the city and place additional burdens on its capacity to assist the poor. Additional taxes on industry, commerce, homeowners, or wage earners to pay for redistributional policies would tend to drive these taxpayers out of the city and reduce the resources available to such policies. According to Peterson (1981: 173), "Broad scale redistributive policy proposals are inappropriately addressed to local governments."

Community elites have an interest in keeping redistributional questions out of local politics, since their resolution may jeopardize members' wealth and income. Elites may manipulate the local policy agenda to exclude redistributional issues, and ensure the "unpolitics" of local poverty, through the now familiar tactics of non-decision-making. In contrast, Peterson (1981) argues that the absence of redistributive issues in local politics "is seldom due to the suppressive activities of an organized economic elite," but rather a product of American federalism and the assignment of responsibility for health and income maintenance programs to state and natural levels of governments.

Nonetheless, redistributive demands are likely to exist in any community, whether these are expressed in local political arenas or not. Class differences are potential fault lines in any community and a latent source of conflict. It is not unreasonable to assert that community elites would prefer to keep these redistributional issues and class conflicts off of the political agenda. Bachrach and Baratz argue convincingly that power is exercised when political institutions limit public consideration to only those issues which do not threaten power holders (Bachrach and Baratz, 1962, 1963). We do not wish to resurrect the theoretical arguments over non-decision-making (Wolfinger, 1971; Merelman, 1968; Debman, 1975). But the theory suggests why community elites might not appear to be actively involved in redistributional politics yet be instrumental in their outcome.

Redistributive demands are occasionally raised in community politics, despite the structure of American federalism and the preferences of community elites (Eisinger, 1973; Lipsky, 1970; Stone, 1976). Riots, protests, rent strikes, demonstrations, and other public manifestations of discontent are threatening to elites. If nothing else, unrest adversely affects the reputation of the community in its efforts to attract new industry. The landed interests cannot altogether ignore redistributive demands.

Community elites have a variety of strategies for dealing with redistributive demands. They may dispense *symbolic* satisfactions in speeches and resolutions, without actually granting any tangible payoffs. They may dispense *token* satisfactions by responding with much publicity to one or more specific cases of injustice while doing little of a broad-based nature to alleviate conditions. Or they may appear to be constrained in their ability to meet redistributive demands by claiming that they lack the financial resources or the legal authority to do anything—the "I would help you if I could but I can't" posture (Schumaker, 1975). Finally, elites can try to *discredit* protestors by stating or implying that they are violent and reckless and unrepresentative of the real aspirations of the people they seek to lead. Most of the empirical research on community conflicts over redistribution suggests that elites generally prevail in redistributive policy issues (see, e.g., Stone, 1976; Lipsky, 1970).

Allocational Policy

Community power structures are seldom directly interested or active in allocational policy. The allocational policy arena is pluralist in character. Decisions about police and fire protection, street maintenance, garbage collection, sewage disposal, school attendance, boundaries, recreation, libraries, public buildings, and the like are usually made by elected public officials or professional managers employed by them. Council members, managers, school board members, and superintendents are responsive to the expressed demands of many varied and often competing groups within the community. Participation in decision making is open to anyone who attends public meetings, gathers petitions, or writes letters to the editor. Interest and activity rather than economic resources are the key to leadership in allocational policy. Access is based upon information about the issues, knowledge of the political processes, and organizational and public relations skills. Community organizations are instruments by which individuals magnify their voice in public affairs. No single group of people dominates decision making in *all* of the service functions of a community's city, county, and school district governments. Elected officials are very sensitive to the opinions of their constituents on service questions. Allocational policies are material and divisable, so allocative policies can easily reflect compromises among competing groups. These are the familiar characteristics of pluralist politics.

Much of what takes place in local government—at commission meetings, public hearings, school board meetings, and in the offices of city managers and school superintendents—involves allocational policies. It is easy for pluralist scholars studying local government to mistake allocational politics for the whole of community politics. This error is facilitated by a narrow and traditional definition of political science as the study of *government,* and the consequent exclusion of community social structure from the purview of pluralist studies. Allocational policies are the "key public decisions" studied by pluralist researchers, and governmental bodies are the locus of their studies. According to Peterson (1981: 150):

> Most of what is discussed as urban politics is the politics of allocation. . . . In this arena political bargaining affects policies,

> and the pattern of bargaining takes a characteristically pluralist form. On allocational issues, one finds acrimonious disputes among those who are at one and the same time united behind developmental policies and uniformly opposed to substantial redistribution. Patterns of coalition formation constantly change as participants find new allies with changing issues. Policy choice is characteristically a compromise among competing interests, the terms of which are influenced by the political leaders' electoral concerns.

Organizational Policy

Organizational policies were the battleground in the struggle between municipal reformers and machine politicians in many of the nation's largest cities for more than a century. Today organizational questions are more likely to concern blacks and Hispanics in their fight to gain representation in local government. The agenda of municipal reform—council-manager government, nonpartisan elections, at-large districts, home rule, merit systems, city planning—was no sooner in place in American cities, when blacks and Hispanics began to challenge some (but not all) of its major organizational accomplishments.

The class and ethnic divisions driving much of the early conflict between machines and reformers has been well described (Hofstadter, 1955; Glazer and Moynihan, 1963). Community power structures are usually only interested bystanders in organizational politics. While it is true that their "class interests" would appear to place them on the side of the upper-class reformers, we know that bankers, developers, and builders had little difficulty in dealing with big city machines. Indeed, machines were often better in "getting things done at City Hall"—that is, moving growth projects to completion, than reformed administrations (Banfield, 1961; Banfield and Wilson, 1963). There is no question that propertied interests thrived under urban political machines.

Community power structures also prosper under reformed administrations. In James Reichley's memorable phrase describing the transition from machine to reform administrations in Philadelphia, the mayor became the business community's "able servant instead of their grafting inefficient slave" (Reichley, 1959: 61). Reformed politicians quickly grasped that good business was good politics, and the most admired city administrations were closely linked to the "economic notables" (Domhoff, 1978). According to Peterson: "Not surprisingly, both business and the city enjoy greater health with an 'able servant' than with an 'inefficient slave' in office. It is thus only an apparent paradox that business prospers most when its influence is least apparent" (Peterson, 1981: 143).

Although machines and reformers differed over which organizational forms should prevail in municipal government and which ethnic groups should get city jobs, they did *not* differ over the goal of economic development. Machine politicians and their reformer protagonists argued over corruption, patronage, and ethnic influences in city government. But they did *not* argue over economic growth or even social distribution.

No doubt community power elites watched these struggles over municipal reform with interest. Perhaps they indirectly aided the reformers in many cities. After all, the reformers were drawn from the same social classes and probably belonged to the same clubs and social circles as the propertied interests. One can imagine upper-class liberal reformers admonishing their friends in business on the evils of bribing machine politicians. But growth-oriented business elites could relax, comfortable in the knowledge that whoever won in organizational politics, economic interests would not be threatened.

Likewise, current disputes over organizational policies—the representation of blacks and other minorities in elected offices and municipal jobs—do not threaten economic interests. While the rhetoric of urban black politics frequently calls for massive income redistribution, the reality of black city administrations is quite different. Whatever fears white propertied elites may have had years ago about black power in the nation's cities should now be placed to rest. Four of the nation's six largest cities (Chicago, Los Angeles, Philadelphia, and Detroit) are now governed by black mayors. There is no evidence that these mayors (Harold Washington, Tom Bradley, Wilson Goode, Coleman Young) or others (for instance, Young of Atlanta, Gibson of Newark, Hatcher of Gary, Barry of Washington, or Movial of New Orleans) are hostile to economic growth and development. Nor have any of them pursued redistributional policies to the point of alienating the business community or driving investment out of their cities. Indeed, the thrust of black redistributional demands—full employment, generous welfare, expanded medical care, publicly financed housing—falls on the federal government.

At the community level, black electoral success has brought no great changes in municipal policies—developmental, distributional, or allocational (Lineberry, 1978; Welch and Karnig, 1979). The only discernible effects of black electoral success have been in ensuring that (1) traditional municipal services are delivered to predominantly black sections of the city (Campbell and Feagin, 1975; Karnig, 1979); and (2) larger numbers and percentages of blacks are employed by the city (Dye and Renick, 1981). Indeed, it is in the area of employment that blacks have been most successful in changing municipal policy. Blacks have been less aggressive in the pursuit of redistributional policies, perhaps because they believe "that the city has less money and controls fewer social services than other governmental levels" (Thompson, 1975: 170). According to Peterson (1981: 161-162), "the responsiveness of governmental institutions to black demands for employment has in fact done much to soften the edges of racial conflict in America. City political leaders have found that responsiveness on issues of employment is a valuable mechanism for incorporating blacks into local political processes without sacrificing the local economy to broad-scale redistribution." Clearly, then, black successes in local politics do not threaten, and may even reassure, community elites about the stability of their investments.

ELITE THEORY, COMMUNITY POWER, AND PUBLIC POLICY

Communities, like societies, are divided into the few who have power and the many who do not. Elite theory poses the central questions of politics: Who has power, how did they get it, and what do they do with it? These questions can direct our inquiries into community politics as well as national politics.

Our argument is that the community's most important resource is land and that those who control the use of land are the community's power elite. This elite centers on the mortgage bankers, real estate developers, builders, and landowners. Community elites are different from national elites in their economic function. The function of community elites is to prepare land for capital investment. But the power of community elites is limited; they cannot fully control the destinies of their communities. The final decisions on where to locate the nation's capital are made by national elites to whom local elites must appeal. Many factors in decisions concerning the location of capital investment are beyond the control of local elites, and elites from different communities must compete for capital investment.

Community elites share a consensus about economic growth. Community elites seek to maximize rents, broadly defined to include appreciation of land values, profit from construction, real estate commissions, mortgage interest, and increased revenues for surrounding service and commercial enterprise. To secure mass support for their policies, community elites hold out the prospect of more jobs and small business opportunities. They argue that capital investment in the community is a benefit to all.

Community controversy arises over the distribution of benefits and costs of economic growth. Growth exacts some costs, and these costs, as well as the benefits of growth, are not distributed evenly to everyone in the community. Persons already settled into the community with secure jobs, persons who own their own homes, persons who will not benefit directly from new investment, and persons who are anxious to preserve their favored life-style form the nucleus of a counterelite. Growth management movements in communities are usually composed of upper-middle-class, educated, articulate homeowners who share an aversion to noise, pollution, congestion, and unsightly development.

Community elites are interested in developmental policies. For the most part, community development is determined by private market forces and not the government. Community elites prefer that developmental decisions be made in the marketplace because of their dominant position in that arena. In contrast, counterelites seek to place developmental decision making in the hands of government, where the economic resources of the propertied interests may be partially offset by the political influence of middle-class homeowners. But most studies of community decision making attest to the prevailing power of community elites in developmental decision making.

In addition to development, communities must make decisions in many other issue areas. We have to choose to construct a typology of community policies, one which helps us to think about the role of community elites in the life of the community.

Our argument is that the role of community elites varies in different types of policymaking. The direct involvement of community elites in policymaking is more likely to occur in developmental policy. The earliest studies of the activities of community power structures appear to describe developmental policymaking. Community elites are likely to employ the tactics of non-decision-making in response to distributional demands. The structure of the American federal system helps to "organize out" of local politics most redistributional questions. Neo-elitist notions about "the other face of power" appear to describe redistributional policymaking in communities.

Pluralist descriptions of community decision making best describe allocational policymaking. Traditionally, pluralist studies focused on *governmental* decision making, ignoring decisions in the private market and overlooking efforts to restructure the agenda of governmental decision making. Local governmental bodies concern themselves primarily with allocational issues: schools, buildings, teacher salaries, police and fire protection, garbage collection, street cleaning, and the like. Allocational policymaking resembles the pluralist model, and pluralist studies of "key public decisions" made by governments provide the empirical support for this model.

Organizational policymaking and the struggle over "reformed" municipal government structures have not been well described by either pluralist or elitist models. The broad-based ethnic and racial divisions in city politics, and the battle for representation in city government, are best described by writers who are not readily categorized as pluralist (Hofstadter, Banfield, Glazer, and the like). But the role of propertied community elites is not easily described either. Elites have prospered under both machine and reformed administrations, and they appear to be prospering today under black mayors as well as they did under their white predecessors.

Community power research has an exciting future. A better understanding of the economic function of community elites provides us with a fascinating agenda for research. Combining this theory with existing typologies of public policy may help us to understand why community research can produce both elitist and pluralist findings.

REFERENCES

AGGER, R., X. GOLDRICH, and B. SWANSON (1964) The Rulers and the Ruled. New York: John Wiley.

BACHRACH, P. and M. S. BARATZ (1962) "Two faces of power." American Political Science Review 56 (December): 947-953.

———(1963) "Decisions and nondecisions." American Political Science Review 57 (September): 632-642.

BANFIELD, E. C. (1961) Political Influence. Glencoe Frec Press.

———and J. Q. WILSON (1963) City Politics. Cambridge: Harvard University Press.

CAMPBELL, D. and J. FEAGIN (1975) "Black politics in the South." Journal of Politics 37 (February): 129-159.

DEBMAN, G. (1975) "Nondecisions and power." American Political Science Review 69 (September): 889-904.

DOMHOFF, G. W. (1978) Who Really Rules? New Haven and Community Power Reexamined. Santa Monica, CA: Goodyear.

———(1983) Who Rules America Now? Englewood Cliffs, NJ: Prentice-Hall.

DYE, T. R. (1976) Who's Running America? Institutional Leadership in the United States. Englewood Cliffs, NJ: Prentice-Hall.

———(1985) Politics in States and Communities (5th ed.). Englewood Cliffs, NJ: Prentice-Hall.

———and J. RENICK (1981) "Political power and city jobs." Social Sciences Quarterly 62 (September): 475-486.

EISINGER, P. K. (1973) "The conditions of protest behavior in American cities." American Political Science Review 67 (March): 11-29.

EULAU, H. et al. (1973) The Labyrinths of Democracy. Indianapolis: Bobbs-Merrill.

GLAZER, N. and D. P. MOYNIHAN (1963) Beyond the Melting Pot. Cambridge: Harvard MIT Press.

HOFSTADTER, R. J. (1955) The Age of Reform. New York: Knopf.

HUNTER, F. (1953) Community Power Structure. Chapel Hill: University of North Carolina Press.

KARNIG, A. (1979) "Black resources and city council representation." Journal of Politics 41 (February): 134-149.

LASSWELL, H. D. and D. LERNER (1952) The Comparative Study of Elites. Stanford: Stanford University Press.

LINEBERRY, R. (1978) "Reform, representation and policy." Social Science Quarterly 50 (June): 173-177.

LIPSKY, M. (1970) Protest in City Politics. Chicago: Rand McNally.

MERELMAN, R. M. (1968) "On the neo-elitist critique of community power." American Political Science Review 62 (June): 451-460.

MILLER, D. C. (1958) "Industry and community power structure." American Sociological Review 23 (February): 9-15.

———and W. H. FARM (1960) Industry, Labor and Community. New York: Harper & Row.

MOLOTCH, H. (1976) "The city as a growth machine." American Journal of Sociology September.

PETERSON, P. E. (1981) City Limits. Chicago: University of Chicago Press.

REICHLEY, J. (1959) The Art of Government: Reform and Organizational Politics in Philadelphia. New York: Fund for the Republic.

SCHUMAKER, P. D. (1975) "Policy responsiveness to group demands." Journal of Politics 37 (May): 488-521.

SILK, L. and M. SILK (1980) The American Establishment. New York: Basic Books.

STONE, C. (1976) Economic Growth and Neighborhood Discontent. Chapel Hill: University of North Carolina Press.

Tax Foundation (1984) Facts and Figures on Government Finance. Washington, DC: Author.

THOMPSON, F. (1975) Personnel Policy in the City. Berkeley: University of California Press.

WELCH, S. and A. K. KARNIG (1979) "The impact of black elected officials on urban social expenditures." Policy Science Journal 7 (Summer): 707-714.

WILDAVSKY, A. (1964) Leadership in a Small Town. New York: Bedminister Press.

WILLIAMS, O. P. and C. R. ADRIAN (1963) Four Cities: A Study of Comparative Policy. Philadelphia: University of Pennsylvania Press.

WOLFINGER, R. (1960) "Reputation and reality in the study of community power." American Sociological Review 25: 636-644.

———(1971) "Nondecisions and the study of local politics." American Political Science Review 65 (December): 1063-1080.

3

The Growth Machine and the Power Elite

A Challenge to Pluralists and Marxists Alike

G. WILLIAM DOMHOFF

> The pedant and the priest have always been the most expert of logicians—and the most diligent disseminators of nonsense and worse. The liberation of the human mind has never been furthered by such learned dunderheads; it has been furthered by gay fellows who heaved dead cats into sanctuaries and then went roistering down the highways of the world, proving to all men that doubt, after all, was safe.
>
> —H. L. Mencken, *Prejudices*

At the sociology meetings around 1975, Nelson Polsby returned to the discipline he had forsaken for the more elevated pursuit of political science. His mission was to tell the sociological "heathen" that the study of community power was a dull and dying, if not dead, field and one that had produced few theoretically interesting results. I think he was right at the time, thanks in no small part to what pluralists had done to the field through their obsessional pursuit of methodological purity. That may be one of the few times Polsby and I have agreed on anything. However, I also like

AUTHOR'S NOTE: I wish to thank Elaine Draper, Beth Ghiloni, Harvey Molotch, and Clarence Stone for their helpful suggestions on earlier drafts of this chapter.

very much what he had to say about both "nondecisions" and "objective interests" in the 1980 revision of his *Community Power and Political Theory*. We do share some common enemies.

To return to the main point, Polsby was right about community power studies in 1975. Not much had happened of any interest in the field since the early 1960s, and we were being subjected to studies of the studies and a lot of back and forth on the same old clichés and canards (see Walton, 1976, for an excellent overview). I know I had not found the field very interesting as far back as 1963, when I first got involved in power structure research. Obviously, I enjoyed Hunter's much misunderstood and unfairly treated book on Atlanta, *Community Power Structure* (1953), although I confess that I never fully appreciated it until I started my own study of community power in New Haven many years later. And, perhaps to the surprise of some, I agreed that Dahl's prizewinning book on New Haven, *Who Governs?* (1961), had demonstrated pluralism for that one city.

However, neither *Community Power Structure* nor *Who Governs?* figured prominently in my own early work. If you take a look at the chapter on local power in my 1967 book, *Who Rules America?*, you will see that it is a brief and perfunctory effort that concedes the local level, at least in New Haven and a few other places, to the kind of pluralistic power conflicts that Dahl portrayed. It then says that community power is not an important issue for those of us interested in "the power elite," meaning the national power structure. I had, of course, borrowed the term from Mills, the other great figure in power structure research (along with Hunter) from whom I take my basic orientation in challenging pluralists and Marxists alike (Ghiloni and Domhoff, 1984; Mills, 1956). At the same time, I had redefined the term to mean the leadership group of the corporate-based national upper class, tying it thereby to a class-based analysis rather than one rooted solely in bureaucratic hierarchies.

I think community power has made its comeback since I wrote *Who Rules America?* and Polsby issued his *pronunciamento*. I believe it is now a very interesting and dynamic field. I am excited to be able to say this, for it did nag at me over the years that I didn't know how to fit the community power studies generated by Hunter and those who followed after him into the research findings on power at the national level. This new work on community power not only ties in with the national-level studies but is closely related to urban sociology and urban political science as well. It provides an underpinning for these two fields and makes it necessary for scholars interested in any urban question—yes, *any* urban question—to take account of community power studies.

I say all these positive and hopeful things on the basis of three general pieces of work. The first is urban sociologist Harvey Molotch's ground-breaking articles, "The City as a Growth Machine," which appeared in the *American Journal of Sociology* in 1976, and "Capital and Neighborhood in the United States," which was published in *Urban Affairs Quarterly* in 1979. The ideas in these articles, and many more, are deepened and defended in further work with another sociologist, John Logan. "Tensions in the Growth Machine" appeared in *Social Problems* in 1984, and *Urban Fortunes* will be published by the University of California Press in 1987.

The second seminal work in the new community power literature is by political scientist Clarence Stone—*Economic Growth and Neighborhood Discontent*. It appeared in 1976 and independently put forth a thesis very similar to Molotch's on the basis of a detailed study of urban renewal decisions in Atlanta between 1954 and 1969. The study was conceived with the controversy between Hunter and Dahl clearly in mind, and it was indeed a masterstroke to apply Dahl's decision-making methodology to the key issue of urban renewal at Hunter's research site.

The third leg of the new community power literature is found in some of my own work. Although I can take no credit for the ideas I present, I am extremely proud of the use I put them to in synthesizing the whole of the community power structure literature in the sixth chapter of my 1983 book, *Who Rules America Now?* There I show how nicely earlier studies of "Middletown," Atlanta, and certain small cities can be illuminated by the new framework.

The chapter in *Who Rules America Now?* also contains the results of my empirical restudy of New Haven in the 1950s, including new findings since my book on the subject, *Who Really Rules?*, appeared in 1978. The book on New Haven presented evidence from memos, minutes, and interview material, including some of Dahl's own interview transcripts that he generously made available to me in the true spirit of scientific inquiry. The book refutes many of the empirical claims Dahl made in *Who Governs?* This includes the all-important claim that the downtown business community was not very interested in urban renewal, making it necessary for the wonderful and brilliant Mayor Richard Lee to "wheedle, cajole, recruit, organize, plan, negotiate, bargain, threaten, reward, and maneuver endlessly to get the support needed from the Notables, the small businessmen, the developers (who came principally from outside New Haven), the federal authorities, and the electorate" (Dahl, 1961: 79). What a man he was, that mayor, and it is no wonder political scientists like Nordlinger (1981: 100-101) continue to quote Dahl's claims as support for their theories about government autonomy without the slightest notion that they lie in a heap of ruins.

My 1978 book is ancient history and out of print, and it did not build on the ideas or findings of Molotch and Stone. The chapter in *Who Rules America Now?* tells the New Haven story deeper and better, and it doesn't hold back in pointed criticism like the 1978 book did when I was still under my long-held delusion that pluralists and Marxists could be engaged in constructive dialogue based on empirical findings.

LOCAL POWER STRUCTURES AS GROWTH MACHINES

So what are these hot new ideas? Drawing from Molotch, Logan, and Stone, the basic argument goes as follows, leaving aside the usual qualifications and refinements that appear in more detailed discussions. A local power structure is an aggregate of land-based interests that make their money from the increasingly intensive use of their land. Contrary to what some of us used to think, it is not a junior-sized edition of the national-level power elite, which is rooted in a nationwide corporate community that sells goods and services for a profit.

In Marx's terms, although virtually no latter-day Marxists use the distinction, local power structures seek after "rents," not profits. The old landlords of the countryside that Marx analyzed are a thing of the European past, but their modern-day counterparts are playing a central role in creating and shaping urban America while the Marxists call everybody with income-producing property a capitalist. As Logan and Molotch (1986: 24) explain:

> Unlike the capitalist, the place entrepreneur's goal is not profit from production, but rent from trapping human activity in place. Besides sale prices and regular payments made by tenants to landlords, we take rent to include, more broadly, outlays made to realtors, mortgage lenders, title companies, and so forth. The people who are involved in generating rent are the investors in land and buildings and the professionals who serve them. We think of them as a special class among the privileged, analogous to the classic "rentiers" of a former age in a modern urban form. Not merely a residue of a disappearing social group, rentiers persist as a dynamic social force.

A local power structure has at its core a set of real estate owners who see their futures as linked because of a common desire to increase the value of their individual parcels. Wishing to avoid any noxious land uses on adjacent parcels that might decrease the value of their properties, they come to believe that working together is to the benefit of each of them. "One sees that one's future is bound to the future of the larger area, that the future enjoyment of financial benefit flowing from a given parcel will derive from the general future of the proximate aggregate of parcels," writes Molotch (1976: 311). "When this occurs," he continues, "there is that 'we feeling' that bespeaks of community."

The most typical way of intensifying land use is growth, which usually expresses itself in a constantly rising population. A successful local elite is one that is able to attract the corporate plants and offices, the defense contracts, the federal and state office buildings, and the educational and research establishments that lead to an expanded work force, and then in turn to an expansion of retail and other commercial activity. Growth also leads to extensive housing development and increased financial activity. It is because this chain of events is at the heart of any developed locality that Molotch calls the urban area and its local elite a "growth machine."

Growth machine. It has a ring to it. It is one of those concepts at once dramatic and insightful that is immediately grasped because it captures so much of our experience and reading. It deserves to take its place alongside "the power elite" as one of the key orienting concepts of power structure research. It is certainly a far richer term than the pluralists' bland and ambiguous "economic notables" or the Marxists' obscurantist and misleading "second circuit of capital."

According to this theory, the most important activity of a community power structure is to provide the right conditions to attract outside investment—or, in Molotch's words, "to prepare the ground for capital." This preparation involves far more than providing level and plentiful acreage with a stream running through it; it also involves the creation of all those factors that make up what is called a "good business climate," such as low business taxes, a good infrastructure of municipal services, vigorous law enforcement, an eager and docile labor force, and a minimum of business regulations. Molotch stresses that the local rentiers expend considerable effort keeping up with the changing place needs of capital. He emphasizes that these rentiers congeal into a middle tier that can be clearly differentiated from both the corporate capitalists and the everyday people. As Molotch (1979: 25) explains the work of the growth machine activists:

> To better understand the needs of capital, and hence to better prepare the ground for them, sophisticated rentiers may take business school courses, read relevant trade journals, make use of their social ties with local capitalists, foster studies at the local university, governmental, or planning agency, or, as is most

> common, use their own "good business sense." The point is that they maintain an attitude of constant alert to the needs of this dominant class.

The growth machines must maintain this attitude of constant alert because they are in competition with each other. The moment a growth machine starts to rest on its laurels, other growth machines will win the new investments. Or maybe even the old investments. For example, if local conditions change in ways that are not to the liking of factory owners, they may move their factories. But land and buildings cannot be moved. This feature differentiates those who produce commodities for profit from those who offer the pseudo-commodity of land as a place for profitable manufacture.

Although the growth machine is based on land ownership, it includes all those interests that profit from the intensification of land use, including developers, contractors, mortgage bankers and related real estate businesses. Executives from the local bank, the savings and loan, the telephone company, the gas and electric company, and the local department store are often quite prominent as well. They too have a strong stake in the growth of the local community. As in the case of the corporate community, where we have great amounts of research on "interlocking directorates," the underlying unity within the growth machine is *most visibly expressed* in the intertwining boards of directors among local companies. (The emphasis is for those followers of French structural Marxists who still do not understand the bourgeois social science distinction between an underlying concept and its many possible objective indicators.) And, as in the corporate community with which we are more familiar, the central meeting points of the growth machine are most often banks, where executives from the utility companies and the department stores join with the largest landlords and developers as members of the boards of directors.

There is one other important component of the local growth machine, and that is the newspaper. The newspaper is deeply committed to local growth so that its circulation and even more importantly its pages of advertising will continue to rise. No better expression of this commitment can be found than a statement by the publisher of the San Jose *Mercury News* in the 1950s. When asked why he had consistently favored development on beautiful orchard lands that had turned San Jose into one of the largest cities in California within a period of two decades, he replied, "Trees do not read newspapers" (Downie, 1974: 112).

The unique feature of the newspaper is that it is not committed to growth on any particular piece of land or in any one area of the city, so it often attains the role of "growth statesman" among any competing interests within the growth machine. Its publisher or editor is deferred to as a voice of reason:

> Competing interests often regard the publisher or editor as a general community leader, as an ombudsman and arbiter of internal bickering, and at times, as an enlightened third party who can restrain the short-term profiteers in the interest of a more stable, long-term, and properly planned growth. The paper becomes the reformist influence, the "voice of the community," restraining the competing subunits, especially the small-scale arriviste "fast-buck artists" among them [Molotch, 1976: 316].

The local growth machine sometimes includes a useful junior partner—the unions of the building trades. These unions see their fate tied to growth in the belief that growth creates jobs. They are often highly visible on the side of the growth machine in battles against environmentalists and neighborhood groups. Although local growth does not create new jobs in the economy as a whole, which is a function of corporate and governmental decisions beyond the province of any single community, it does determine where the new jobs will be located. Thus local unions find it in their interest to help the growth machine in its competition with other localities.

Jobs, of course, are something the landowners like to talk about too. Jobs are the ideal unifying theme for bringing the whole community together behind just about any growth project. The goal of the growth machine is never money-making, but jobs for the citizens of the community:

> Perhaps the key ideological prop for the growth machine, especially in terms of sustaining support from the working-class majority, is the claim that growth "makes jobs." This claim is aggressively promulgated by developers, builders, and chambers of commerce; it becomes part of the statesman talk of editorialists and political officials. Such people do not speak of growth as useful to profits—rather, they speak of it as necessary for making jobs [Molotch, 1976: 320].

Although a concern with growth in general is hypothesized by Molotch to be at the basis of each local elite, every city enters into the competition with a different set of priorities and strategies for achieving it. Calculations have to be made about what investment possibilities are the most desirable and possible based upon such factors as the availability of natural resources; the nature of the climate; the proximity of oceans, lakes, and rivers; the skills of the work force; and the past history of successes and failures in growth competition. Obviously there is a clear preference for clean industries that require highly paid skilled workers over dirty industries that use unskilled workers, but dirty industries will be accepted if other locales win the clean ones. Attractive beachfront towns are not as likely to seek out just any type of industry, as are inland cities; their property can bring in more money as sites for tourist resorts and convention centers. When an area has little or nothing to offer, as in the case of most of Nevada, it settles for gambling and prostitution to create a Las Vegas. Growth strategies can also change over time; when Atlantic City lost out as a "nice" resort, it adopted the Nevada strategy and turned to legalized gambling.

Historical factors also enter into growth strategies. If one locale gets there first with a once-in-a-generation opportunity, such as a stockyard or a railroad, over which competition was very fierce in the nineteenth century, then nearby communities have to settle for lesser opportunities even though they have very similar natural conditions. On the other hand, earlier successes may lock an area into relationships and obligations that make it very difficult for it to take advantage of new opportunities. The rise of Sunbelt cities is not only due to cheaper labor costs, the availability of land, and good weather, but also to the previous lack of heavy industry that made growth machine leaders there more alert to the possibilities in electronics and information processing.

The growth machine hypothesis leads to certain expectations about the relationship between power structures and local government. Obviously, the primary role of government is to promote growth according to this view. "It is not the only function of government," writes Molotch (1976: 313), "but it is the key one and ironically the one most ignored." Local government promotes growth in several ways, the most visible of which are the construction of the necessary streets, sewers, and other public improvements and the provision of the proper municipal services. However, government funds for the boosterism that gives the city name recognition and an image of togetherness, are also considered important by the growth machine in attracting industry. Sometimes the money for boosterism is given directly to the Chamber of Commerce. In some places, it is given to an Industrial Development Commission or a Convention or Visitors Bureau that is jointly funded by government and private enterprise. Then too government officials are expected to be the growth machine's ambassadors to outside investors, traveling to meet with them in their home cities or showing them the local community and answering their questions when they come to inspect it for possible investment.

Because so many specific government decisions can affect land values and growth potentialities, leaders of the growth machine are prime participants in local government. Their involvement is even greater than that of corporate capitalists at the national level, where the power elite can rely to some extent on such "signals" as stock prices, interest rates, and the level of new investments to tell government officials what they think of current policies. The growth machine is the most overrepresented group on local city councils, as numerous studies show, and it is also well represented on planning commissions, zoning boards, water boards, and parking authorities, the decision making bodies of greatest importance to it. However, this direct involvement in government is usually not the first or only contact with government for members of the growth machine. They often have previous service on the local Chamber of Commerce's committees and commissions that are concerned with growth, planning, roadways, and off-street parking (Daland, 1956; Bouma, 1962; Kammerer et al., 1963; Lowry, 1964; Prewitt, 1970).

But the growth machine does not operate without strains, tensions, and conflicts, and it does not dominate local government without constant effort. Local landowners and developers do battle with each other for advantages on occasion, and I have already pointed out that rival cities vie for the needed capital investments from outside. Then too there can be tensions with the capitalists who invest in the city, as the great concern over plant closings most recently attests. As conditions change, such as the increasing international division of production, capital can move elsewhere, leaving behind a broken or stalled growth machine.

Equally important, there is sometimes conflict between the growth machine and specific neighborhoods. Neighborhoods are something to be used and enjoyed in the eyes of those who live in them. However, they are possible sites of further development for the "highest and best use" in the eyes of those who run the growth machine. This conflict between "use value" and "exchange value" is a basic one in most successful cities as the downtown interests try to expand. It is not a mere by-product or displacement of the capital-labor conflict so captivating to most Marxists that they cannot see what is really going on in city politics. Sometimes the neighborhoods win, especially when they are aided by organized environmentalists or supplemented by university students. In little Santa Cruz, California, for example, the growth machine has not won a single battle on freeways or major new developments since students on the local campus of the University of California began voting in 1972.

The nature of the battle between growth machines and neighborhoods should not be overstated. Neighborhoods are usually stable and apolitical, and they very often lose when they try to fight change. As Logan and Molotch (1986: 46) note:

> Ordinarily there is relative peace. Community groups usually mobilize only when their own particular turf is threatened. Lacking overall interest in the use of the metropolis as a whole, they are not a source of ongoing resistance. And because of relative weakness, urban community organizations are always vulnerable to cooptation by those with larger and more stable resources.

It is the conflict between the growth machine and the neighborhoods that is revealed in fine detail in Stone's book on urban renewal in Atlanta. Simply put, he found that land values and growth were the key issues in the city's politics. Urban renewal was only "part of a general struggle over the control of land" (Stone, 1976: 45). Stone came to this conclusion because he found that urban renewal in Atlanta was based on the desire to expand the central business district into land occupied by low-income black neighborhoods that also were in the process of expanding.

The growth machine's concern for urban renewal was expressed through the Central Atlanta Improvement Association. Members of this group involved themselves with government through membership on the City Housing Authority and a Citizens' Action Committee on Urban Renewal that was jointly financed by business and government. There were also informal contacts with both of the mayors who served during this era, the first a lawyer and longtime ally of business interests, the second a downtown businessman and former president of the Chamber of Commerce.

Responsibility for specific government plans was lodged in an Urban Renewal Planning Committee composed of commissioners from the Housing Authority and elected officials from the city council's Committee on Urban Renewal. Unlike what Dahl found in New Haven, the mayor stayed in the background as much as possible. Responsibility for the program was insulated from him so that he would not become the center of any political controversies over it.

Stone studied several specific decisions from both the formative years of the program, 1954-1962, and the more protest-laden years of 1962-1969. He found that the downtown business community was overwhelmingly successful in achieving its major objectives during the first phase. Its only setback was a partial one of little direct interest to it: Sites for low-income housing that it had agreed to as the price for black leadership support were blocked by a middle-income white neighborhood (Stone, 1976: 82-83).

In the second phase the downtown business interests were supportive of high-rise office buildings, a sports complex, and the expansion of a downtown university while quietly vetoing the repeated requests of specific neighborhoods for public housing. This second phase was highlighted by black demonstrations and protests. These actions should have met with great success if pluralistic theorists are right that "the prizes go to the interested and active" (Stone, 1976: 90). For a time, it did look as if they were going to be successful. In early 1966 a neighborhood group was able to save one-half of its area from plans for urban renewal clearance. The outbreak of relatively mild civil disorder in two other black neighborhoods in September 1966 galvanized the city into promises for larger recreational and fix-up programs for neighborhoods. In November 1966 the mayor announced a goal of 17,000 new units of low- and moderate-income housing, with 9800 of those units to be completed within two years. By the end of 1966, Stone (1976: 128-129) reports that the program seemed to be changed rather dramatically:

> The change had come abruptly. As late as the 1965 Declaration of Policy in the city's Workable Program document, Atlanta's urban renewal program was explained primarily in terms of the "encouragement of economic expansion," physical planning and development, and "the overall economic ability of the City to support . . . urban development and renewal activities." By the close of 1966 the urban renewal program was completely recast; neighborhood improvements, grass-roots participation, and expanded supply of standard housing for low- and moderate-income families appeared to be central elements in a new renewal policy.

However, the policies did not change after all. As protest receded, the promises went unfulfilled. City officials stalled and delayed as they fought with neighborhood groups and as the groups within neighborhoods began to argue with one another. Only a few thousand of the 17,000 promised housing units were built. City officials blamed neighborhood opposition for this failure. Stone (1976: 150) suggests that lack of business support was even more important, for the business leaders had made it clear that they preferred low-income housing to be built outside the city limits. The opposition of real estate interests to any government involvement in the construction and management of housing also went unmentioned by city officials.

Stone sees the results of his decisional study (and please note that this is the type of study pluralists advocate) as a direct contradiction of the general theory of pluralist politics that Dahl derived from New Haven. Poor citizens do not have what Dahl called the "slack," or extra resources, to invest in politics when they feel their interests are threatened enough to make politics worthwhile. Despite sustained protests and other efforts, blacks in Atlanta were not able to induce politicians and government officials to forward their interests. Instead, the link between the growth machine and city hall—based on common values, organizational ties, and campaign finance—proved more durable. Between 1956 and 1966, one-seventh of the people in Atlanta were moved out of their homes to make way for expressways, urban renewal, and a downtown building boom that drew nationwide attention throughout the 1970s (Stone, 1976: 3, 227).

This viewpoint also encompasses my findings on New Haven, findings that contradict Dahl's claim that Mayor Lee and his aides had to interest the downtown business community and Yale in urban renewal, among other claims. In fact, the business community and Yale had been central to the development of urban renewal plans that complemented their growth plans. I found minutes and memos which show that leaders in the local business community met with Lee within two weeks after his election in November 1953 to urge their program on him. Yale archives reveal that Yale and the city had been negotiating on land for Yale since October 1951—two years before Yale and the city made contact according to the standard sources. I will not quote again from the documents that support these claims beyond the shadow of a doubt, but I will say that they deserve to be read and pondered by those who keep footnoting the conclusions in *Who Governs?* as support for this or that theory as if they had never been challenged empirically.

The new scheme also works well on Hunter's findings on which policy issues interested Atlanta's leaders. Contrary to the canard that Hunter was not interested in policy questions in his interviews, Hunter in fact asked his respondents, "What are the two major issues or projects before the community today?"—which, by the way, happens to be a much better way to pick issues for further study than the arbitrary way Dahl and other pluralists do it. Of 26 top leaders in the growth machine, 23 gave the plan for growth created by the Central Atlanta Improvement Association as one of their two choices. Second, with nine votes, was a plan developed by the business community's Traffic and Safety Council to move traffic in and out of the city more quickly by means of new highways. No other issue received even five mentions (Hunter, 1953: 214-215). Twenty years later, in 1970, when Hunter returned for a second study and asked the same question, he received the same answer. The number one interest was in physical improvements related to growth, but this time it was a rapid transit system, a new airport, and a downtown sports and convention center. "In the interviews, they could speak of nothing else," reports Hunter (1980: 150-152).

By way of comparison, Hunter found very different emphases in other parts of the community. The 14 professional workers he asked the same question of in 1950 gave as many votes (four) to housing and slum improvement as they did to the plan of development, and 3 of the 14 also mentioned improved race relations. The contrast with black leaders was even more striking: blacks thought that improved schools and better housing were the main issues, and they placed better housing and higher employment as their top concerns in 1970.

The new theory also demolishes the alleged "refutation" of Hunter's study published in 1964 by political scientist M. Kent Jennings under the title *Community Influentials: The Elites of*

Atlanta. According to Polsby (1980: 147n, 161), this study shows that Hunter was wrong in this emphasis on the power of "economic dominants," and Jennings has let that judgment stand in a fulsome review of Polsby's book for the *American Political Science Review* (1981: 211-212).

But of what does this great refutation consist? Mostly of the fact that the local manufacturers and the branch plant managers of national corporations, said by Jennings to be "economic dominants," are not named by knowledgeable informants as the most powerful people in Atlanta. Jennings has tapped into the fact that industrialists and their plant managers are not highly involved in the growth machine. This type of finding had been reported in earlier studies as well, leading to a pseudo-controversy over possible changes in local power structures due to the rise of the "absentee ownership" of large corporations. The problem largely disappears if we understand the distinction between the power elite and the growth machine.

What Jennings actually found in Atlanta is a complete vindication of Hunter and his dreaded "reputational method," not to mention a boost for the idea of a growth machine. The fact that this simple truth has never been acknowledged is a commentary on the power of pluralism within the political science fraternity. When the main study comes up with 57 of the 59 names uncovered in the pilot study, and when those names include 23 of the top 27 from Hunter's study of several years earlier, then there is evidence that the method is a lot better than anyone later admitted (Jennings, 1964: 24, 25, 156). This is especially the case when it is added that Jennings found many of these people were involved in the issue areas that he studied. As Jennings himself finally wrote (1964: 164), many pages after he presented his findings:

> From a methodological point of view, our findings show that the nomination-attribution technique is neither so infallible as its supporters claim nor so misleading as its attackers insist. Most of the perceived influentials at both levels were indeed influential in one or more issue areas. Those considered most influential tended

> to engage in more deliberately influential behavior and appeared actually to be more influential than those reputed to be less influential. The technique measures more than simply respect, popularity, or social status. It serves to locate people of consequence in community decision making.

Of course, as I have already hinted, the findings turn out to be supportive of the growth machine argument. Even though most of the "perceived influentials" were not "economic dominants" by Jennings's definition, the majority were businesspeople who came by and large from the downtown business interests: "Those who were economic dominants as well as perceived influentials came mainly from nonmanufacturing firms, which are located in the central business district, are locally owned, and have their locus of consumption in the immediate metropolitan area" (1964: 197).[1]

Perhaps you can begin to see why I think the growth machine argument is a theory with considerable potential. It even explains empirical findings.

PLURALISTS AND MARXISTS

As I said at the outset, I am very excited by this new way of thinking about community power structures and their relations with each other and national-level power. It is an extremely dynamic theory that deals with all the conflict that so dazzles pluralists but at the same time remains true to the findings that we have on Atlanta, New Haven, other communities, and national power. It leaves me with a feeling of closure about the past community power structure literature. It also gives me hope that some members of the next generation of historians and social scientists will find these ideas worth trying out on new issues and research sites. This is especially the case when we consider the pluralist and Marxist alternatives.

What do the pluralists have to offer? As far as I can tell, they are still arguing about methods or trying to count up the number of faces that power wears—is it one, two, or three (Lukes, 1974).[2] Indeed, there seems to be only one pluralist-type book on the city that anyone talks about: political scientist Paul Peterson's award-winning *City Limits*, published in 1981.

Peterson improves on previous pluralist studies by mentioning that "Urban politics is above all the politics of land use," arguing that cities "compete with one another so as to maximize their economic position" (1981: 25, 28). After some gratuitous comments (1981: 136) about Hunter doing research in a "casual manner" and writing in a way that "fluctuated from sociological obscurantism to journalistic sensationalism," thereby ensuring in advance that no one will accuse him of embracing elite theory, Peterson tells his fellow pluralists that there is more to Hunter's work than has been previously realized. Even with all his alleged failings, Hunter uncovered those who are involved in the politics of development. In one sense these people are a relatively insulated elite, but they are not the cause of development politics. Instead, development is due to the competition among cities, which is just the opposite of the growth machine formulation.

Since development is something everyone can agree on, it is just good politics. There is no need for any growth machine to work up boosterism, and of course Peterson has no theory concerning the potential conflict between the growth machine and neighborhoods. Naturally, Molotch and Stone are not among the many works that he finds worth using. Although he notes the limits of extreme pluralist formations, he writes about the initiative of Mayor Lee in New Haven as if that claim never had been challenged empirically. If this is the best pluralists can do, then pluralism simply is not alive and well for anyone interested in understanding what is happening in American cities.

As for the Marxists, they are so trapped in their most abstract and militant-sounding categories that when they look at urban areas they can see nothing but "accumulation," "circuits of capital," "class struggle," and "the reproduction of labor power." Their first major mistake is that they generally collapse the important distinction between capitalists and the place entrepreneurs of the growth machine. Even when they see differences they do not understand that the landlords have played a local and intermediary role. Instead, they make distinctions between "fractions of capital," and at best they see these "fractions" differing over profits in housing versus lower wages for workers. As Walton (1981: 384) reports in a review of the new urban sociology, several leading Marxists "have analyzed the manner in which expenditures on collective consumption (housing being prototypical) tend to divide fractions of capital, with those dependent on land, construction, or rental favouring greater profit in housing while industrial capital opposes such upward pressure on wages."

The same error of collapsing categories also takes place on the other side of the great dialectical divide, claiming thereby that people are only workers and that all conflicts are class struggles. Marxists fail to distinguish the defense of neighborhood use values from class struggles over exploitation and surplus value. Thus Tabb and Sawers (1978: 5), writing the introduction to their edited volume on *Marxism and the Metropolis*, tell us that in the 1960s "struggles over urban space intensified as community groups fought for their homes against highways and urban renewal." Then they leap to the claim that "many came to see these struggles over 'turf' as forms of class struggle." They do not tell us exactly who these "many" are who came to see things this way, but from a reading of the subsequent essays, it is clear that the category encompasses all the Marxists in their volume who speak in general theory.

For urban geographer David Harvey, one of the two major figures among Marxian urbanists, the problems of the city are crudely termed "mere reflections of the underlying tension between capital and labor." The whole reductive quote reads as follows:

> Conflicts in the living space are, we can conclude, mere reflections of the underlying tension between capital and labor. Appropriators and the construction faction mediate the forms of conflict—they stand between capital and labor and thereby shield the real source of tension from view. The surface appearance of conflicts around the built environment—the struggles against the landlord or against urban renewal—conceals a hidden essence that

> is nothing more than the struggle between capital and labor [Harvey, 1976: 289].

It is hard to beat this quote for pure abstractness, or to miss the implication that all but Marxists are fooled by surface appearances and the deceits of capital. Everything, and hence nothing, is explained by class struggle. Marxism is truly what Mills called it—a labor metaphysic.

The other leading urban Marxist, Manuel Castells, does not make the mistake of reducing urban social conflict to an aspect of the class struggle. Furthermore, he understands that the conflict between exchange values and use values is a basic one in the city; he criticizes fellow Marxists because they "have tended to respond [to rival theories] by reducing the city and the space to the logic of capital" (Castells, 1983: 297). Unfortunately, though, he goes off the deep end in a completely opposite direction, labeling every social movement that happens to occur in an urban area an "urban social movement." Despite the fact that most of the case studies he presents in his well-known *The City and the Grassroots* (1983) were in fact battles between landlords and tenants that accomplished very little, or else conflicts over government urban renewal projects, he claims that these and other urban-based movements (gay rights, women's movements, black uprisings) are aspects of a much wider and more profound struggle that has three underlying goals: "collective consumption trade unionism" (that is, urban amenities and housing), the search for cultural identity ("community"), and the decentralization of state power to city governments and neighborhoods. These goals in turn relate to the three themes that dominate recent world history:

> The three goals that are crucial factors in the fulfillment of urban social movements are precisely the three alternative projects to the modes of production and modes of development that dominate our world. The city as a use value contradicts the capitalist form of the city as exchange value. The city as a communication network opposes the one-way information flow characteristic of the infor-

> mational mode of development. And the city as a political entity of free self-management opposes the reliance on the centralized state as an instrument of authoritarianism and a threat of totalitarianism. Thus the fundamental themes and debates of our history are actually the raw material of the urban movements [Castells, 1983: 326].

Here the conflicts between the growth machine and specific neighborhoods disappear into an abstract Neo-Marxian logic that is as grandiose in its own way as Harvey's orthodox class-struggle logic. The city and its power structure are lost from sight once again. Castells's effort is a noble and heartfelt one, but as Molotch (1984: 141-143) writes in a review of the book, Romantic Marxism is not good enough either.

In truth, neither pluralism nor Marxism is good enough. Their original proponents contributed important ideas, but neither set of epigone has been able to use the ideas from the rival camp to go on to bigger and better things. Fortunately, the growth machine perspective accomplishes the needed synthesis. It preserves the mainstream emphasis on markets, which is an essential starting point, but it shows how markets are sociologically constructed and explains how land markets differ in important ways from commodity markets. The theory resurrects Marx's distinction between capitalists and landlords that has been ignored by latter-day Marxians. It also uses the Marxist distinction between exchange value and use value to understand the conflict between growth machines and neighborhoods, while at the same time emphasizing that these conflicts are analytically distinct from struggles between capitalists and workers. Thus it is a new theory because it is a creative synthesis of the best available ideas from both schools of thought. And it is a good theory because it encompasses existing empirical findings and immediately suggests hypotheses for further studies.

It is time to make a new start in community power studies. It is time to get pedantic pluralists and priestly Marxists alike out of their sanctuaries and into the fresh air of new ideas.

NOTES

1. It is no small task to figure out the composition of the group of "attributed" or "perceived" influentials in the Jennings study (1964). Table 1 (p. 46) reports that 30% are in commerce, 15% in finance, and 15% in manufacturing, so at least 60% are businesspeople. Another 34% are in a category called "professional and public administration," but we also learn that 30% of the perceived elite had legal training. This information, along with that on page 52, suggests that some of those who are legally trained may be corporate lawyers. Finally, 6% of the perceived influentials are listed as retired, but Jennings shows remarkably little interest in their former careers. We learn that there are two blacks, one a lawyer who runs the black voters' league and the other the president of a black-run financial institution (p. 39). We are then told (p. 40) that there is one woman in the group, the superintendent of public schools. On page 120 we learn that one influential is a labor leader. We learn that the executive director of the business-dominated Central Atlanta Association is a perceived influential (p. 102), but we do not know if he is a businessman or a professional. It is my hypothesis that the great majority of the perceived influentials are part of the growth machine, but I have been unable to obtain a list of their names.

2. It is my view that all the participants in the sterile and tiresome debate over the "faces" of power, including the "radical" Steven Lukes himself, are pluralists.

REFERENCES

BOUMA, D. (1962) "Analysis of social power position of a real estate board." Social Problems 14 (Fall): 121-132.

CASTELLS, M. (1983) The City and the Grassroots. Berkeley: University of California Press.

DAHL, R. (1961) Who Governs? New Haven: Yale University Press.

DALAND, R. (1956) Dixie City: A Portrait of Political Leadership. Birmingham: University of Alabama, Bureau of Public Administration.

DOMHOFF, G. W. (1967) Who Rules America? Englewood Cliffs, NJ: Prentice-Hall.

———(1978) Who Really Rules? New Haven and Community Power Reexamined. New Brunswick, NJ: Transaction.

———(1983) Who Rules America Now? Englewood Cliffs, NJ: Prentice-Hall.

DOWNIE, L. (1974) Mortgage on America. New York: Praeger.

GHILONI, B. and G. W. DOMHOFF (1984) "Power structure research in America and the promise of democracy." Revue Française D'Etudes Americaines 21/22 (November): 335-341.

HARVEY, D. (1976) "Labor, capital and class struggle around the built environment in advanced capitalist countries." Politics and Society 6: 265-295.

HUNTER, F. (1953) Community Power Structure. Chapel Hill: University of North Carolina Press.

———(1980) Community Power Succession. Chapel Hill: University of North Carolina Press.

JENNINGS, M. (1964) Community Influentials: The Elites of Atlanta. New York: Free Press.

———(1981) "Book review: Community Power and Political Theory, by Nelson Polsby." American Political Science Review 75 (March): 211-212.

KAMMERER, G., C. FARRIS, J. DEGROVE, and A. CLUBOK (1963) The Urban Political Community. Boston: Houghton Mifflin.

LOGAN, J. and H. MOLOTCH (1984) "Tensions in the growth machine." Social Problems 31: 483-499.

———(1986) "Urban fortunes." University of California, Santa Barbara. (unpublished)

LOWRY, R. (1964) "Leadership interaction in group consciousness and social change." Pacific Sociological Review 7 (Spring): 22-29.

MENCKEN, H. L. (1924) Prejudices: Fourth Series. New York: Knopf.

MILLS, C. W. (1956) The Power Elite. New York: Oxford University Press.

MOLOTCH, H. (1976) "The city as a growth machine." American Journal of Sociology 82 (September): 309-330.

———(1979) "Capital and neighborhood in the United States." Urban Affairs Quarterly 14 (March): 289-312.

———(1984) "Romantic Marxism: love is still not enough." Contemporary Sociology 13 (March): 141-143.

———and J. LOGAN (1984) "Tensions in the growth machine: overcoming resistance to value-free development." Social Problems 31: 483-499.

NORDLINGER, E. (1981) On the Autonomy of the Democratic State. Cambridge: Harvard University Press.

PETERSON, P. (1981) City Limits. Chicago: University of Chicago Press.

POLSBY, N. (1980) Community Power and Political Theory (2nd ed.). New Haven: Yale University Press.

PREWITT, K. (1970) The Recruitment of Political Leaders. Indianapolis: Bobbs-Merrill.

STONE, C. (1976) Economic Growth and Neighborhood Discontent. Chapel Hill: University of North Carolina Press.

TABB, W. and L. SAWERS [eds.] (1978) Marxism and the Metropolis. New York: Oxford University Press.

WALTON, J. (1976) "Community power and the retreat from politics: full circle after twenty years?" Social Problems 23: 292-303.

———(1981) "The new urban sociology." International Social Science Journal 33: 374-390.

4

Power and Social Complexity

CLARENCE N. STONE

Students of power have long debated the question of whether it is possible for a small group in a modern democracy to exercise broad control over society. This chapter is an attempt to displace rather than resolve that debate. Put another way, I shall try to reformulate the issue. The word "control" is the point we need most to rethink. In all likelihood we should simply abandon it, but then we are left with the question of how to think about power. I wish to pose the issue this way: what aspects of power are appropriate for analyzing relationships in complex social systems? If control, as conventionally understood, is inappropriate, what term or terms are needed?

I have posed the issue in reference to complex social systems because I think we have a special need to think about what it means politically to live under conditions of social complexity. To the extent that this essay makes an original contribution, it will be less about power per se than about the special character of power relationships in complex social systems.

This essay will be largely a twofold exercise, first in defining terms (but not in the highly formal sense) and second in sketching some possible relationships based on those terms. My main concern will be to develop some key orienting concepts and to show that they can lead to theoretical reformulation. In proceeding from definitional exercise to theoretical reformulation, I will talk descriptively about how power operates in systems that are constitutionally democratic but socially complex at the same time. I will draw in particular on examples from the city of Atlanta, which is the political entity I know in most detail. However, I want to emphasize that this discussion is intended to be illustrative and not definitive. My purpose is to develop a theoretical orientation and not to attempt to provide confirming evidence. The latter can come only if I can succeed first in making the case that the theoretical orientation offered here is worth pursuing. The examples, then, are not evidence in a strict sense, but simply an attempt to make concrete what otherwise might be too abstract.

Let me offer one final prefatory note. The power terms I employ—command power, bargaining power, intercursive power, and ecological power—are only partly familiar. "Ecological power," in particular, is a term I have coined in this chapter expressly to capture what is distinctive about power relationships in socially complex systems. Yet almost everything said in this chapter owes a heavy debt to the pioneering efforts of others. The particular usage and theoretical formulation presented may be somewhat unique to this context, but let me begin with general acknowledgments to obviate the need to sort out step by step the intellectual debts I owe others. The term "ecological power" was inspired largely by Long's (1958) provocative article, "The Local Community as an Ecology of Games." As a form of power, the term also draws heavily on the extensive work associated with Schattschneider's (1960) notion of the mobilization of bias, Bachrach and Baratz's (1970) concept of non-decision-making, and the contributions of several other authors to what is often known as the second and third faces of power (Anton, 1963; Crenson, 1971; Gaventa, 1980; Lukes, 1974). In addition, the work of Lowi (1979) and McConnell (1966) on the private capture of public authority is highly instructive background for the line of argument developed here. Additional specific debts are acknowledged in the usual manner.

Some readers may wonder why, in the discussion that follows, I have not used the term "systemic power" that was central in previous work I have done on power (Stone, 1980). Let me offer a brief response to that question, one that foreshadows the overall argument I want to offer. This present essay is an effort to cut into the problem of power from a different angle. There is a widely held belief that complexity fragments power, and in some respects it does. However, in other respects it does not, and we lack the conceptual language to differentiate. The earlier conception of systemic power is helpful in moving away from a Weberian vocabulary, one based on a concern with the power to command. But the notion of systemic power does not go far enough. It focuses on the features of a stratified system that make some groups more attractive as policy partners, but the notion of systemic power is purely passive. It says nothing about what the attractive policy partner may do in order to make use of and even expand its systemic advantage. The concept of ecological power, as used in the following, is an attempt to address that point. It is therefore an extension, not an abandonment, of the notion of systemic power.

DIRECT AND INDIRECT POWER RELATIONSHIPS

The prevailing notion of power centers on direct control. In its purest form, control or domination consists of a capacity to issue commands that cannot be successfully resisted. According to Weber's influential formulation, power is "the possibility of imposing one's will upon the behavior of others, even against opposition" (see Bendix, 1960: 290). Bachrach and Baratz (1970: 27) offer the closely parallel statement that "in a power relationship one party obtains another's compliance." And Wrong (1980: 2) defines power as "the capacity of some persons to produce intended and foreseen effects on others."

In a recent book on democratic theory, Dahl largely forsakes the term "power" for its synonym "control." He defines the latter as "a relation among actors such that the preferences, desires, or intentions of one or more actors brings about conforming actions, or predispositions to act, of one or more other actors" (Dahl, 1982: 16). Dahl introduces the notion of influence through the intermediary of the social structure and makes the provocative observation: "Much of what people do is influenced by the restraints and opportunities provided by social structures" (1982: 18). He declines, however, to call that a form of power.

The linking of power to direct control, from Weber to the present, has important implications. Pluralist conclusions are embedded in this conception of power. Wrong's book on power (1980) illustrates the subject. Following de Jouvenel's earlier lead, Wrong (1980: 14) distinguishes three attributes of power relations:

(1) extensiveness—the domain or number of power subjects;
(2) comprehensiveness—the scope or variety of actions over which the power holder can move the power subjects to compliance; and
(3) intensity—how far the power subjects can be pushed without loss of compliance.

Wrong distinguishes these three attributes in order to make a key theoretical point about built-in limitations to control. The greater the domain (extensiveness) of a power relation, the more limited the scope of actions (comprehensiveness) and the less the intensity of the control are likely to be. Wrong (1980: 20) explains:

> There are three main reasons why the greater extensiveness of a power relation sets limits to its comprehensiveness and intensity. First, the greater the number of power subjects, the greater the difficulty of supervising all of their activities. Second, the greater the number of power subjects, the more extended and differentiated the chain of command necessary to control them, creating new subordinate centers of power that can be played off against each other and that may themselves become foci of opposition to the integral power holder. Third, the greater the number of subjects, the greater the likelihood of wide variations in their attitudes toward the power-holder. The power-holder will not be able to wield power with equal comprehensiveness and intensity over all his subjects [see also Banfield, 1961].

Domination is thus rarely complete. Even in total institutions, forms of bargaining arise between inmates and their overseers

(Goffman, 1961; Sykes, 1958). As Dahl (1982: 33) points out, subjects may be able to establish a degree of autonomy by raising the costs of domination. After all, he observes, "control is almost always to some extent costly to the ruler" (Dahl, 1982: 33).

Given the difficulty of maintaining domination, especially over a varied population, we would expect to find that control in a complex society operates only in narrow domains. Hence, it may be argued that modern society is not characterized by a ruling class that exercises comprehensive control. Instead, elites with narrow scopes of control find that they must bargain with one another (see especially Keller, 1963). Each controls too narrow a range of activities to have much effect alone.

Wrong also puts forward another conception of power—what he terms "intercursive power" which he contrasts with integral power. The latter refers to monolithic control by one element. As Wrong (1980: 11) explains, integral power is "centralized and monopolized by one party." By contrast, intercursive power is associated with pluralist bargaining; it is "characterized by a balance of power and a division of scopes between the parties" (Wrong, 1980: 11). Consider, for example, the Chicago depicted by Banfield (1961) in which Mayor Daley, with his command of land-use controls, bargained with business leaders who controlled credit and investment monies and with leaders in the construction trades who controlled the skilled labor supply. This constitutes pluralist bargaining among leaders who command narrow domains of activity, and this has been the prevailing view of power in American society. It is not an inaccurate view so much as an incomplete view. Dahl (1982: 193) directs our attention to the missing element when he talks of "the present system of decentralized bargaining by exclusive associations, which are strongly motivated to pass on the cost of their bargains to others not involved in the bargaining."

The actions pursued by individuals have consequences, albeit often indirect, for others. This is to say that behavior of one actor may alter the environment in which another actor operates. Of course, the consequences of strictly individual actions are likely to be small. The actions pursued by coalitions of actors also have consequences for others, but if the coalition is (to use the key word) powerful, the consequences may be quite substantial—although indirect (they are not consequences based on direct control of the affected population). A coalition that can modify the social environment by altering the terms of interaction or modifying structural arrangements in some way is indeed powerful. Those who live within the environment of these interactions and arrangements (what is called here "the social ecology") but who cannot alter their environment have good cause to feel powerless. Those who can make modifications will do so in ongoing efforts to protect and promote their interests, as Dahl observes, pushing the side effects and indirect costs onto others. Those who cannot make modifications are likely to find their interests neglected at best and in some instances adversely affected. Thus conflicting interests may be at work, not through a direct struggle between contenders, but by means of unequally equipped contenders attempting to reshape the conditions under which they operate—that is, to reshape the social ecology. Let us call this last category of relationships an indirect power relationship.

We are now in a position to talk about different dimensions of power. First and most familiar, there are exercises of power under conditions of direct conflict. If the relationship is sufficiently asymmetrical for one party to dominate, this is clearly a case of *command power.* If the actors are in direct conflict over some objective, but neither has sufficient power to dominate, then they may bargain. In this first dimension, the difference between dominating and bargaining is the degree to which the power relationship is one-sided. Our definitional focus, however, is on the strength of the power of command, hence we will use the term command power for this first dimension.

There is a second dimension of power that involves bargaining, mentioned above as *intercursive power.* This is the power associated with coalition formation. Actors with complementary resources, perhaps complementary domains in which each has command power, negotiate in order to agree on terms of cooperation. This cooperation enables them to undertake a large task collectively that they would have been unable to pursue individually, or to solve a problem collectively that they were unable to solve individually. Intercursive power is enabling power—"power to," as contrasted with dominating power, or "power over," as exemplified in command power.

In thinking about coalition formation, we need to note that there is not only a direct power relationship—the agreement to cooperate. There is also a potential indirect power relationship—the spillover effects of action by the coalition on parties not members of the coalition. Because of the spillover effects, coalition activity can, under some circumstances, engender opposition from a challenging coalition. In this instance, the contending coalitions become engaged in a "power over" struggle that may result in one coalition dominating or in the contending coalitions bargaining over a compromise. Nevertheless, the process of coalition formation—the exercise of intercursive power—is itself different from any conflicts the coalition may be drawn into.

At this point, it is important to bear in mind the generally acknowledged principle that control is costly. If domination involves an ongoing effort to gain compliance, the struggle may not be worthwhile. That is why a materially poor but spirited and determined opponent is not easily dominated. Thus political actors, including coalition representatives, generally try to avoid opposition, sometimes by institutionalizing the advantages they seek. For example, actors may try to secure an advantage through a favorable structuring of the terms under which social interaction occurs (that is, to institutionalize practice). Union officials are strong advocates of seniority rules. Black leaders seek affirmative action plans. Once rules are institutionalized, they become part of the established order and may be hard to challenge. Of course, rules vary in the degree to which they are visible to and likely to be challenged by others. Some—affirmative action plans, for example—are conspicuously redistributive and therefore evoke considerable opposition. Others—tax advantages for historic preservation, for example—are so obscure and minutely redistributive in their immediate impact that they are unlikely to evoke much opposition.

It is no easy matter to gain the institution of rules favorably structuring the terms of social interaction, but the obstacles to be overcome do not always come in the form of opposition from other parties. Instead, as will be spelled out later, the barriers may take the form of a need for special capacities: (1) being master of the kind of information needed to formulate rules that are technical enough in content and indirect enough in other-party impact to avoid opposition, and (2) being able to secure the cooperation, usually from government officials, needed to put rules into practice.

This capacity to secure the institution of favorable rules of social transaction has no recognized label, and I am going to designate it *ecological power*. This is our third dimension of power. As the discussion up to this point indicates, ecological power is an amalgam of already recognized forms of power. It grows out of control over resources (including in some instances the loyalty of able subjects) and coalition membership. It thus rests on a base of command power and intercursive power. I am using a separate label for it as shorthand, particularly as shorthand to suggest how, under some conditions, a limited amount of direct control of strategically important resources can be parlayed into the capacity to achieve an outcome that has an indirect effect over a large domain.

Ecological power gains in importance to the extent that people are related to one another in complicated and indirect ways. Power thus has dimensions other than that of direct control, and the problem of power is more complicated than that of compliance versus resistance. The term ecological power serves to remind us that power superiority can consist of something other than the capacity to command obedience. It can also consist of a superior power to reshape the social environment. Therefore, I wish to define ecological power as the capacity to reshape the context—that is, the social ecology—within which one operates. The strength of ecological power increases to the extent that it is not counteracted by competing demands from other power actors also seeking to reshape the social ecology.

It is important to remember that a superior capacity to achieve modifications in the social ecology is still a limited capacity. It is not a capacity to reshape at will. However, it is also not a power that is equally shared. For that reason, some parties are in a position to gain advantages in pursuing their interests while others suffer disadvantages. The advantage or disadvantage from any given modification may be modest, but repeated adjustments yield cumulative advantages and disadvantages that are likely to be quite substantial (on cumulative advantages and disadvantages, see Stone, 1976).

Let me emphasize a related point. There is a logical difference between having a superior capacity to pursue a goal, on the one hand, and, on the other hand, having the capacity to impose on someone a result to which that party is opposed. The latter would be a case not only of "power to" but of "power over" as well. Ecological power relationships involve the possibility of this circumstance of the imposition of unwanted results. As one party achieves a modification of social arrangements and conditions of social interaction, that changed social ecology has effects that may be detrimental to the interests of other parties. However, the imposition is indirect—that is, it is mediated through sufficient steps to avoid a direct conflict. Because the effects are mediated through third parties, arise out of multiple causes, or are delayed in time, they are hard to trace. Hence, the imposition may not be readily attributed to any one set of events or actors. In actual research practice, it is much easier to identify the differential capacity to reshape the social ecology (the superior capacity to pursue a goal) than to establish the degree to which a given ecological effort by one group had an adverse impact on the interests of others. Put another way, it is easier to look at the competitive capacity to reshape the social ecology than it is to look at the imposition of particular unwanted effects.

At this point the reader is almost certainly asking why we should want to look at power and politics in such a convoluted way. Is it worthwhile? Two related points should be kept in mind. First, so long as we think about power in terms of direct exchanges, we tend to be caught up theoretically in issues about the costs and difficulty of achieving compliance over resisting subjects. In this way, analyses of power resources and tactics come to center on part of the picture, leaving out other parts. Second, other parts of the picture are increasing in importance. The nature of modern society is such that indirect power relationships are of heightened consequence. To understand why this is so, and to see how this circumstance affects the practice of politics, particularly urban politics, we turn now to a brief discussion of social complexity.

SOCIAL COMPLEXITY

In one sense, all but the most primitive societies are complex. Of course, there are matters of degree. What I wish to address here are advanced complex systems characterized by a high level of interdependence (see especially Brewer, 1973; LaPorte, 1975). Moreover, interdependence is of a particular kind. Because interdependence is so extensive, any given activity tends to have consequences throughout the system. Reciprocal relationships are numerous, and interactive results are common. There is what Downs (1967: 156) calls "an extremely complicated web of relationships." Results are filtered through several actors and steps. Consequences may be delayed, particularly if played out through numerous reactions and interactions.

To appreciate the political implications of social complexity, consider how transactions are conducted in simpler as compared to more complex systems. In a relatively simple system, relationships are direct, and market-like exchanges work well. The consequences of exchanges are largely confined to the direct parties in the negotiations. In contrast, complex systems are characterized by long-linked relationships that are mediated through several parties (see Thompson, 1967). Further, the consequences of any given activity are not easily confined (economists use the terms "externalities"; others speak of spillover effects) and tend to ramify throughout the system. An adversely affected party may have no immediate recourse. Moreover, because interdependence is so extensive and relationships so complicated, lines of causation are not easy to trace. Most significant outcomes have multiple contributors and can be altered only by substantial and multifaceted efforts. Several factors may contribute to an outcome, or an outcome may result from the accretion of individual acts—any one of which would itself be inconsequential (see Hardin, 1968).

Given the inadequacy of direct negotiation and exchange for solving problems, dissatisfied parties (the well-off at least as much as the have-nots) turn to government for help. However, given our uncertainty about the chains of causation, government cannot serve as the arbiter of disputes. Instead, government is drawn heavily into problem-solving efforts, often as the source of compensatory assistance. Because relationships are complicated, most problems are not thought of in terms of dispute resolution. And even where there is a dispute to be resolved, resolution may be complicated by spillover effects on third parties, which may themselves have to be attended to. As complexity asserts itself, government becomes less visible as an arbiter and more visible as a mobilizer and coordinator of resources. Programs that result may in turn have consequences which can trigger proposals for additional problem-solving efforts (see Wildavsky, 1979).

Complex society is very much a system of indirect relationships. Direct and mutually enforced agreements encompass too few contributing pressures to be effective in bringing about desired change. Because only government combines the enforcement authority and the revenue capacity needed for genuinely broad-scale efforts, those seeking to alter the social ecology in which they operate tend to work through government. Frequently, those promoting change in the social ecology attempt to bring about public-private partnerships (for examples of business interest in such arrangements, see Fosler and Berger, 1982). Of course, those working for public-private partnerships may also seek to limit participation in the partnership and insulate its activities from the general body of popular pressures. The capacity to secure the use of governmental authority and resources is thus at the heart of ecological power. It is in varying capacities to enlist government that differences in ecological power become evident.

As is the case with any other type of activity, governmental programs have direct as well as indirect consequences. Direct consequences that have substantial adverse effects (urban renewal and urban expressway construction are obvious examples) engender strong reactions from those whose interests are harmed, and programs are likely to be abandoned or modified to forestall intensifying opposition. (The costs of achieving compliance from a resisting population are part of this pattern.) However, many effects are so diffuse (e.g., the costliness of rapid rail transit and its detrimental impact on bus service) or so delayed and indirect (e.g., the impact of high-rise development in raising property values and altering street life to the detriment of profit margins for small shopkeepers) that they elicit little countervailing opposition. That there are many such adverse effects is certain, but these effects are not easily pinpointed, even in elaborate analyses. Certainly diffuse, delayed, and indirect effects provide weak ground for political mobilization. Complex systems thus contain no built-in assurance of mutual adjustment, despite the fact that they might be democratic and provide avenues for popular pressures.

This is not to suggest that there are no instances of the pluralist scenario of bargaining between directly affected parties, mediated by public officials. Rather, it is to suggest that the pluralist bargaining scenario captures only a small part of political life in socially complex systems. Politics under conditions of complex causation is more a matter of competing priorities—whose public-private partnership will prevail and on what terms—than of mutual bargaining. This is what ecological power is about, the question of who has the competitive edge in being able to enlist government in restructuring the terms under which social interactions occur.

It might be objected that diffuse, delayed, and indirect consequences are nothing new. Even machine politics of an earlier era, with its extensive reliance on relationships of direct exchange, were nevertheless a form of politics with enormous indirect consequences and displaced costs. That point I fully concede. What distinguishes the current era is not the presence of such consequences but the extent to which there is a body of political participants who are mindful of these consequences and who act accordingly. What distinguishes the contemporary era, then, is an awareness of social complexity and the capacity by some participants to act purposefully in guiding or deflecting its consequences. Ecological power is about an unevenly distributed capacity to cope with complexity.

ECOLOGICAL POWER IN OPERATION

Before looking further at ecological power, it might be useful to take stock of the argument developed so far. First, I have suggested—following the lead of Wrong and others—that the nature of power does not lend itself to the establishment of direct and intense control over a large domain in a wide scope of activity. Because such control is costly to maintain, large and diverse societies are likely to be characterized by narrow domains of command power that may be connected in various ways by bargaining to establish coalitions for the control of larger domains. These expanded bases of power may be used to alter structural arrangements—that is, to modify the terms under which social interaction occurs. This extended form of power—ecological power—is best thought of as a capacity to act, as power to do something rather than as power over a body of subjects. Yet it rests in command over resources, and this control over resources puts some actors in the position of being able to bargain and extend the range of activities subject to power efforts. Even so, ecological power is not a direct form of control and is not a capacity to reshape society at will. It does, however, involve modifying the social environment in which a large domain of subjects operates. It is in that sense an indirect, albeit limited, form of power.

Since ecological power can be used to protect and advance the interests of some to the detriment of the interests of others, we need to ask why such exercises of ecological power are not resisted. Why does ecological power not result in a direct struggle and thereby come to be indistinguishable from command and bargaining power? Sometimes it does, but often it does not. The nature of social complexity explains why it often does not. Social complexity does not reduce to direct bargaining relationships. The affected activities are not so easily segmented and identified. Mediated and interactive effects, delayed effects, and effects that come from accretion make it hard to pinpoint causes. Further, such complicated interactions make it hard to identify the principal parties in a transaction, much less bring them together for a bargaining exchange.

With direct, face-to-face bargaining unavailable as a response to many unwanted conditions, government serves as a frequently invoked problem-solver. Moreover, the complicated causation that makes many problems highly intractable calls for the extensive and sustained mobilization of resources. With its extensive authority and revenue capacity, government is an especially attractive partner in efforts to either modify the conditions that give rise to problems or to ameliorate the consequences of the problems. Hence, ecological power is typically exercised by enlisting government as a partner in a problem-solving venture. As I shall argue in more detail, the business-government relationship is not one of domination. Rather, it is one of mutual attraction, of filling complementary needs. The formation of the partnership is an example of intercursive power. The use of the partnership is where ecological power comes into play. Intercursive power is thus a vital link to ecological power.

If ecological power were equally available to everyone, we might enjoy some complicated form of democratic utopia. But the nature of ecological power is such that it is not easily accessible to everyone, and its consequences are not open to resistance in the same way that direct forms of control are. Resistance, for example, cannot take the form of passive and partial (and therefore hard to detect) noncompliance. It can only take the form of total withdrawal from society's ongoing processes, which would be effective only if done on a massive scale. That would require a difficult and, in terms of resources expended, costly effort. Scattered acts of "resistance" cost the wielder of ecological power little or nothing. Indirect power can be checked only by a countermobilization. For direct power relationships, it is evident that control can be costly. For indirect power relationships, the opposite is the case. Under these conditions, costs fall very heavily on those who would resist because resistance can occur only through an organized effort. Indirect power relationships thus have a dynamic that is quite different from direct control. That is the reason for this exercise in trying to develop a new perspective and a new vocabulary to describe that perspective.

We are ready now to move beyond our stocktaking of what ecological power is and why it enjoys special prominence under conditions of social complexity. At this point, we need to consider what characteristics make some actors more effective than others as exercisers of ecological power. Two characteristics seem especially appropriate:

(1) possession of strategic knowledge of social transactions and a capacity to act on the basis of that knowledge; and
(2) control of resources that make one an attractive coalition partner, and in particular to public officials.

It might be argued that almost everyone is at least vaguely aware of the importance of general structures of social transaction. To take some obvious examples, most people understand that it matters whether a given activity is handled through market exchange, the efforts of volunteers, or a public agency. They are also likely to believe that elected and appointed school boards respond to different sets of cues. But they may be totally unaware of the rules, structures of incentives, and unwritten norms that surround most activities. While they have a general idea about market exchange versus public authority as transaction processes, they may not know how tax laws work to tilt the profit advantage from rental housing toward condominium conversion. They may be aware of food stamps but not know about the tax advantages of historic preservation districts. They probably know the term "old boy network," but they may have little knowledge of most informal circles of mutual assistance. Or they may have a clear understanding of cliques in the offices where they work but know nothing of the relationships between corporate boards and corporate chief executive officers.

Much knowledge is vague, and that is the point—that it is vague. Most people are aware that corporate executives enjoy perquisites, but few know what a "golden parachute" is. Even when detailed information becomes available, it may not become part of an understanding of how the world works on a continuing basis. Kickbacks, payola, pension fund abuses, and various forms of private and public corruption make the news occasionally but are not matters of continuing public concern. Illegal transactions are an important but small part of the universe of social transactions. Many forms of cooperation rest on tacit understandings and are fully legal but can be known readily only by those who are part of the immediate system of social transactions.

Information about forms of social transaction is not scarce, but systematic, reliable, and detailed information is costly. It can be acquired only by great effort, and few individuals, *as* individuals, acquire such information. And in a society of large-scale operations, few individuals are in a position to make much use of such information even if they had it. Information about schemes of incentives and disincentives, and about how systems of cooperation are related to one another, is sought and used by large, formal organizations. These are structured to search out such information and fit it into their operations. The first thing that we can expect about ecological power, then, is that such organizations will play a prominent role. They have planning and analytical capacities (on the national importance of these capacities, see Dye, 1983; Domhoff, 1983). In the power terminology I am using, we can say that large, formal organizations are in exceptionally good positions to be informed about social ecology and to act on that knowledge.

Unorganized and loosely organized groups are unlikely to have concrete and usable knowledge of social ecology in any detail. Their behavior in public affairs is often reactive and consists of responses to acute crises. They rarely have broad and systematic plans of action for restructuring segments of the social environment. Consumers are less likely than producers to be effectively organized. While union officials are likely to have considerable information about social ecology, they may be unable to communicate that information to rank-and-file workers in such a way as to elicit desired forms of mass action (for example, union officials are often unable to deliver voting support). Racial, religious, and ethnic groups—all mass membership groups, for that matter—have the same problem of how to achieve disciplined collective action in support of plans formulated by a leadership hierarchy.

At the community level, not only do individual businesses have units devoted to planning and analysis, but they are often the only interest organized into peak associations. Chambers of commerce and improvement associations for the central business district are invariably key actors at the community level, and their involvement is directed toward ecological power. The reason they exist is to analyze the portions of the social environment most closely related to business activity, develop plans based on that analysis, and launch efforts to enlist member participation in the policies that result from analysis and planning. Neighborhood and residential associations, on the other hand, have little planning and analysis capacity. They are oriented mainly toward reactive responses to particular threats. In the terms I am using here, they are not well positioned to engage in comprehensive analysis and planning as a basis for efforts to seek ecological reordering.

The notion of a social ecology is sufficiently abstract and remote from most daily concerns as to be outside the operating frame of reference of most participants in politics. With the exception of large and formally organized groups, few do more than react to specific events and seek remedies for particular problems. In the local community, business stands almost alone as the one interest organized in such a way as to be knowledgeable about and concerned on a continuing basis with something so abstract as the finely graded variations in terms of which social transactions occur. Similarly, at the local level it is business that is primarily concerned with how different systems of transaction are linked together.

In a complex society, this capacity to be informed and to act on information about the social ecology is enormously important. To be without this capacity is somewhat like playing chess as if the strategic objective were to capture the pawn your opponent moved last. To play effectively, one needs an overall sense of how various pieces relate to strategic control of the board. One cannot anticipate every move, but one is lost without thinking several possible moves ahead and making moves within a general strategy of play. In a complex system, whether it is chess or community economic development, knowledge is power—certainly a first and vital step toward exercising ecological power.

The second element in enjoying an ecological power advantage has to do with attractiveness as a coalition partner. Solving or—perhaps more accurately—ameliorating a problem under conditions of social complexity is no easy task; it is likely to require a sustained and multifaceted effort. That is why government is attractive as a partner. But consider the partnership from the other direction (see Stone, 1980). Whom are public officials likely to find attractive as members of a public-private coalition?

Bear in mind that we are talking about coalitions engaged in reordering the terms of social transaction in efforts to solve or ameliorate problems. Clearly, there is no shortage of problems—public officials have considerable freedom to decide which problems will receive priority. To any given demand, numerous responses are possible. Officials can deny that they have a capacity to act; they can provide a token or short-term response;

they may treat it as a matter for symbolic action and engage mainly in posturing; or they may make a substantial and sustained effort and thereby enter into a prolonged partnership with the concerned private party (on the strategy of various responses to protest, see Lipsky, 1970). In deciding which response to make, officials undoubtedly consider many factors. But the nature of complexity assures that a consideration will be the resources commanded by private interests—those they can bring to a problem-solving partnership.

Efforts aimed at ameliorating largely intractable problems provide few immediate payoffs of the kind appealing to a mass electorate. Many such efforts engender as much popular opposition as popular support. In any event, mass appeal tends to be short-lived. Popular sentiment, then, is not a very useful guide in selecting which problems to make the target of a long-term undertaking.

Problem-solving activities increase in attractiveness as the private partner is able to contribute from its own store of resources and to commit itelf to a sustained effort. A private partner with more resources will have greater attractiveness than one with fewer resources, but amount is not the only factor. Other things being equal, a partner with a variety of resources is more attractive than one with a narrow repertoire of resources. And a partner with indispensable resources—that is, needed resources that are not readily found elsewhere—is especially attractive (on this latter point, see Emerson, 1962; on resources and urban coalitions, see Stone, Whelan, and Murin, 1986). At the community level, business is likely to have the edge on all these counts.

Social complexity heightens the attractiveness of yet another feature of resource possession. Because problem-solving efforts require time, a group's attractiveness as a coalition partner increases with the durability of its ability to provide resources. Resources based on temporary surges of enthusiasm among a diffuse group membership or mobilization that hinges on a crisis atmosphere are not of much use for long-term program efforts.

Contemporary government is executive government, and it works mainly through bureaucratic forms of organization, as do businesses. Bureaucracies represent durable resource mobilizations and find it easiest to work with one another. Bureaucracies also seek a stable environment (see Thomspon, 1967). As bureaucratic organizations attempt to achieve such an environment, they find it easier to negotiate with a single large association, the behavior of which is predictable, than with numerous small entities, the collective behavior of which is uncoordinated and unpredictable. By the nature of things, unity is more likely among a few large entities than among many small entities, especially if the few are geographically concentrated enough to enjoy substantial interaction and the many are too dispersed for anything more than limited interaction (see Olson, 1971; Hardin, 1982).

In terms of individual actors and their perspectives, bureaucratically based operators prefer to deal with other bureaucratically based operators as they attempt to establish ongoing arrangements of cooperation. It is thus easier for public officials to work with a chamber of commerce than with an aggregation of neighborhood associations. The key is not business over nonbusiness, however, but bureaucratic and therefore predictable over nonbureaucratic and therefore unpredictable. A single large business, such as a bank or major department store, is likely to be predictable and sufficiently resource-laden as to be worth negotiating with itself. Even better in some respects is negotiation with a peak association. Not only does the peak association represent a wider domain of supporting resources, it also manages the problem of coordination and internal unity among members. That is a major reason why it is easier to work with a chamber of commerce than with neighborhood organizations. Or, to confine the comparison to groups within the general category of business, that is why it is easier to work with a downtown improvement association than with, for example, the owners and managers of rental properties (on this comparison, see the article on the post-home rule experience in Washington, D.C.; *Washington Post*, December 31,

1984). Both are business groups, but one is capable of unified action and long-term commitment while the other is not. To the extent that social complexity heightens the importance of arrangements that are predictable and long-lasting, it also heightens the attractiveness of alliances with business associations and individual large firms.

Overall, the characteristics of social complexity serve to elevate the importance of knowledge about how society operates, especially in a detailed and operational sense. It also makes especially attractive as allies those groups that control substantial resources, especially if their resources can be drawn on in a durable and predictable relationship. Because control of deliverable and needed resources makes a group attractive to public officials as a partner in a problem-solving venture, the dependable control of resources becomes a major stepping-stone to ecological power.

In some ways the ability to exercise ecological power is a competitive matter. Those groups with better strategic information have an edge over those with inferior information, and those groups better endowed with resources enjoy an advantage over those with lesser endowments. However, there is more than a competitive edge at work. Because the informational and resource demands are considerable, ecological power is not readily available to everyone. Information and resource capacities must exceed a threshold in order for a group to be effective. Not everyone can enter the competition, and at the community level, where organizational capacities for most groups are limited, there are few contenders. Although I have stated the argument largely in general terms, it should be evident that major business interests most easily meet the criteria of effectiveness, and at the community level they may well enjoy a near monopoly.

APPLICATION OF THE ARGUMENT

The argument about ecological power and social complexity is intended to reorient the way in which we think about power, particularly in the community setting. In the conclusion I suggest that, to the extent that such terms as "ruling class" and "ruling elite" imply direct forms of control, they are inappropriate for understanding power in modern society. At the same time, the notion of ecological power enables us to see how certain forms of direct control restricted to fairly narrow domains of activity can nevertheless be used to achieve significant power advantages that are indirect in nature. We need to move away from the notion that power necessarily entails a relationship of direct struggle between domination and resistance. The image of one party that commands and another that complies or resists is deeply embedded in some definitions of power. I suggest a different image, one of competitive advantage in reshaping the social ecology.

The notion of ecological power also has implications for how we think about the modern state and its role. Particularly in a time when there is much laissez-faire rhetoric, it is useful to keep in mind just how important a role the state plays in setting and adjusting the terms under which social transactions, including market transactions, occur. Before tackling the broad theoretical ramifications of the argument, we should consider how it might apply to concrete events and activities. These concrete events and activities help point us toward future directions for research.

Much of the work in power has come out of or been tested in community studies. Power at the community level is the particular concern of this book, and I will therefore direct my discussion to the city experience.

Numerous efforts have been made to account for the preponderant influence of business in city politics, especially in development policy. Those labeled elitists have pointed to the seeming capacity of business interests to control the agenda of public debate and decision making (Lynd and Lynd, 1937; Hunter, 1953; Bachrach and Baratz, 1970; Domhoff, 1983). Others have concentrated on the role of money in politics (Caro, 1974) or on the particular needs that business fulfills. Clark (1969), for example, cites the low prestige of city politicians in America and the capacity of business leaders to provide civic legitimacy.

Peterson (1981) highlights the economic competition among local jurisdictions and maintains that Main Street and City Hall, along with the populace generally, have a unitary interest in economic growth. Neo-Marxists make a similar point in talking about capital accumulation as a system need in capitalist political economies (Whitt, 1982; Fainstein and others, 1983; Tabb and Sawers, 1984). None of these arguments is without supporting evidence, but all address governmental arrangements in a fairly general way.

The ecological power argument addresses governmental arrangements directly. Consider how the conduct of city affairs has changed as society has become more complex and, even more importantly, as some participants have come to recognize the complexity of modern life. In earlier times, when direct exchanges were prevalent, machine politics held sway. Transactions occurred between individuals who headed personal networks of supporters or personally run enterprises. Patronage, the disclosure of inside information, and other forms of personal favoritism were the glue that held things together. In contemporary times, recognized by some as being more complex, the scale of activities and the inadequacy of direct exchanges have helped eclipse machine politics. Arrangements whereby favorable treatment is bestowed on those who are part of the governing coalition have become more intricate and indirect.

Cities and states create special purpose agencies, often independent of direct governmental control, to handle a range of development policy activities. Either formally or informally, business is usually given special representation in these arrangements. Atlanta, for example, has formed an economic development corporation headed by a board representing business interests and local government. Significantly, the Atlanta Economic Development Corporation is housed not in City Hall but in a private office building in quarters adjoining the local chamber of commerce (on the Atlanta experience, see Stone, 1984; on general patterns of local government and private business cooperation, see Fosler and Berger, 1982).

Land redevelopment is frequently turned over to quasi-public corporations that are business-dominated. To use Atlanta as an example, I cite the case of a large tract of land east of the central business district, land acquired by the city through eminent domain. In this instance, Central Atlanta Progress, the peak association of downtown businesses, was allowed to form a nonprofit corporation, Park Central, to control the disposition of land for development. Turning over responsibility to a private group served, Park Central maintained, to signal that redevelopment was not to be "some sort of social experiment" (Henson and King, 1982: 324). A similar arrangement has been proposed for the redevelopment of Underground Atlanta. Another independent authority handles the provision of tax-free development bonds, still another the Atlanta Stadium, and yet another the rapid transit system. In each case board appointees are predominantly businesspeople. Further, the injunction to follow business-like practices and avoid social experiments holds for all of these special purpose agencies. In Atlanta, there are a number of publicly funded and publicly backed agencies through which major development transactions occur, and these are largely insulated from popular pressures. To appreciate the full extent to which these arrangements represent a restructured social ecology, one needs to bear in mind that with the 1970 census, Atlanta joined the list of black-majority cities, and blacks have controlled City Hall from 1974 to the present. Protecting development activities from popular pressures is thus a significant altering of the political terms under which transactions occur.

More is at issue than formal organizational arrangements, although these are important in their own right. Businesses throughout urban America have been able to negotiate an extensive set of subsidies: tax-exempt bonds, tax-increment districts (through which enhanced revenues from development are kept separate from general revenues and reserved for additional improvements in the development district), tax abatements, written-down land costs (often of land acquired through the public power of eminent domain), publicly built and publicly underwritten convention and exposition centers, subsidies and guarantees for professional sports, rapid depreciation allowances for office buildings and other real estate developments, and tax advantages from historic preservation designations (see Jones and Bachelor, 1985). In addition to various forms of direct financial subsidy from all levels of government (many of which do not show in government budgets at any level because they are tax rather than revenue expenditures—others show in the capital but not the operating budget), business interests have been able to arrange for public facilities to be located so as to complement and sometimes stimulate private development plans (on Atlanta, see Stone, 1984). In addition, business districts sometimes receive special intensive patrolling by police (see Cingranelli, 1981) or special "clean-up" campaigns by law enforcement authorities aimed at changing the "character" of an area.

Special organizational arrangements, subsidies, and informal agreements to cooperate in support of development efforts are all manifestations of ecological power—of the capacity of business interests to structure the terms under which major transactions occur, but to do so in a way that forestalls participation by other community interests. In what way, it might be asked, are we justified in linking these events to the term "ecological power"? Certainly they are examples of what business influence can achieve. But why "ecological"?

These are not examples of individual favors being handed out to particular actors, in the manner of patronage politics. Instead, they are forms of benefit that come through defining the terms under which transactions occur (in that sense, restructuring the social ecology). Though business efforts are usually more successful, business has no exclusive monopoly on such efforts. Neighborhoods in Atlanta, for example, succeeded in having a Neighborhood Planning Division created in the City Planning Department, but that venture in a city-neighborhood partnership proved to be short-lived. Minority set-asides and affirmative action plans also represent efforts at restructuring and have had some impact. However, they are now under legal and political attack. What is unique about business efforts to achieve restructuring is the extent to which business interests succeed in this area without close public scrutiny. They are able, with reasonable ease, to shift back and forth between market and nonmarket criteria and even to mix the criteria.

Business uses its position as a partner in development policy to alter the terms under which the shift between market and nonmarket criteria can occur. On some occasions it is argued that interference with market processes will have unwanted consequences and is sure to leave everyone worse off through inefficiency. On other occasions, it is argued that government should play a major role (in land acquisition, for example) because governmental authority can be used to overcome the fragmentation of the market. It may be further argued that market processes are inattentive to the sunk costs and long-term investments in the central business district, and that only government planning and assistance can preserve these essential community assets. Thus many development activities are extensively subsidized and may even depend on publicly acquired land (land acquired by other than market processes). Yet once the development activity is under way, social claims are forbidden and market criteria are declared supreme (for a parallel, non-Atlanta example, see Weinberg, 1977: 147-176). Then, too, many of the subsidies are so indirect, so technical, or, in other ways, so hidden from public view that there is little likelihood of a public clamor for nonbusiness criteria to be used.

CONCLUSION

The concept of ecological power serves to direct discussion away from the conventional forms of power analysis. Consider the two dominant views. The left-critical view has centered on the notion of ruling elites. According to this school of thought, economic and governmental powers are conjoined in a system of elite domination. Coercion and the capacity to promote legitimating beliefs provide ruling elites with the means through which they maintain control. Although no elitist maintains that a ruling body exercises tight control, most elitists believe that there is a more or less cohesive circle of power figures who rule in the sense of maintaining control of the agenda of public decision making.

In contrast, the centrist and largely celebratory view of American democracy depicts a system in which governmental and economic power are largely separate, and in which governmental officials are meaningfully accountable to popular sentiment via elections. In this view elites are unable to maintain unity, and they cannot put into effect a scheme of control. According to this school of thought, there are too many divisions among elites and too much opportunity for popular resistance for elites to exercise mass control. As a result, bargaining and exercises of countervailing power are common. Those adversely affected by a policy have the resources and are motivated to mobilize and bring about change. Hence, mutual adjustment and pluralist accommodation are frequent outcomes.

The so-called elitist/pluralist debate has rarely navigated between the dangers of each side. On one side is the Scylla of extraordinary business influence. Its existence can hardly be denied. On the other side is the Charybdis of insufficient evidence of unified elite control. How do we reconcile the condition of fragmented control with the condition of highly unequal influence? Neither conventional argument offers a widely persuasive answer.

The argument developed here steers our attention toward indirect power relationships—the essence of ecological power. We need to see that the fragmented system of control (that is, narrow domains of command power) is only a part of the overall power picture. We also need to see that, in a liberal democracy, the state is an important participant in power relations even though its coercive power is sparsely used, has a restricted application when invoked, and, at least in America, has a low intensity. The Weberian understanding of the state as an instrument of coercive control perhaps highlights the wrong features of modern government practice. The state is not only a police officer, it is also a societal "traffic manager"—determining what forms of social transaction will be used and how they will be related to one another.

It is important to realize that in a system of fragmented control, where overtly coercive power is used sparingly, concerted actions are nevertheless possible, though not uniformly available to all. The possession of strategic information and the ability to organize in order to act on strategic information vary. Groups differ enormously in the kind and quantity of resources they control and, of at least equal importance, in the manner in which they control and deliver resources. While most groups can react defensively against some immediate threat, few combine information, organization, and resources in such a way as to be able to wield ecological power. Few cross the threshold of effectiveness in being able to form the kind of alliances needed to restructure the terms under which social transactions occur. Even fewer can do this in such a way as to be unchallenged in the restructuring they seek.

Social complexity means that everyday observations and ordinary intuitions are not enough; complexity puts a premium on the possession of strategic information about how the social ecology works. Further, there are rules of coalescence that carry special weight under conditions of complexity. Actors seek partners who are most useful—typically those who have substantial resources, complementary and essential resources, and deliverable resources.

In urban communities the threshold of effectiveness is sufficiently high to keep most groups from exercising ecological power. Hence, few groups have the capacity to modify the terms under which social transactions occur. In socially complex systems, no one issues a comprehensive set of commands to the mass citizenry, but coalitions of actors do negotiate the terms under which the mass citizenry operates.

The complexity of relationships stands in the way of the spontaneous development of countervailing power. In a simpler system, where the source of adverse effects could be clearly identified, opposition to those adverse effects and their perpetrator might be easily organized. At least, this would likely be the case where no dominant group held control. In a complex system, however, different circumstances prevail. The actions of a policy coalition might have adverse effects, but such effects may be mediated through several steps, may be diffuse, or may be delayed and even somewhat concealed. If the effects are mediated, reactive responses may be off-target. If the consequences are diffuse, opposition will be hard to organize. If the effects are delayed or partially concealed, counterpressure may come too late to prevent the offending actions from becoming entrenched and encased in supporting alliances. In the full flowering of complexity, consequences would be mediated, diffused, and delayed—thus making effective opposition triply unlikely to be mobilized.

Bear in mind that the policy coalition responsible for adverse effects on others may have to do nothing to prevent a countervailing coalition. What makes the ecological power relationship indirect is that the power-wielder is effective without having to engage in direct actions to prevail over opposition or to prevent opposition from developing. If potentially rival coalitions are too dispersed to achieve effective collective action or are divided by racial or religious identities, the effective policy coalition holds a competitive advantage whether it had to work for such an advantage or simply enjoys it from the nature of the circumstances. The strength of a competitive advantage does not depend on how hard the advantaged worked to gain that advantage. Put another way, a superior *power to* act can be possessed without exercising a power of prevention *over* others. Nevertheless, a superior power to act can (and, under condition of social complexity, does) have spillover effects on others. What makes it superior power is that others cannot counter the spillover effects and have little or no capacity themselves to impose effects on others. One or more parties are making adjustments in the social ecology to suit their interests; others are not—they are nonparticipants. The pluralist "principle of affected interests"—that any group substantially affected by an activity has the opportunity to participate in the governance of that activity—breaks down.

Groups differ in the degree of their awareness of being affected and in their capacity to act on those things of which they are aware.

To the extent that a superior power to act evokes no countervailing response, it is part of a set of indirect power relationships. That is the form that a competitive edge in restructuring social ecology often takes. By blurring the consequences of actions, complexity contributes to the development of indirect power relationships. Under these conditions, the diversity of interests has a noteworthy effect: It contributes to the difficulty of forming countervailing coalitions. As the consequences of a policy action are dispersed over a varied set of interests, perhaps affecting each in a different way, counter-organization becomes difficult. Thus, whereas exercising direct control over a diverse body might be difficult or at least costly, an indirect power advantage is most easily maintained over a diverse body of competitors. Without direct control, there is no unifying focal point for grievances. Under conditions of complexity, disappointments are often blamed on circumstances—and in an immediate sense, circumstances are indeed the culprit. People do get caught up in the imperatives of situations (see especially Schelling, 1971). What distinguishes the ecologically powerful from the powerless is the capacity to understand and alter the situation. While no one has complete understanding or full capacity to remake circumstances, some have a greater understanding and capacity than others, and they dominate the process of ecological readjustment.

Because indirect consequences are prevalent, even the role of the state may not be fully comprehended. Under conditions of social complexity, the democratic state occupies a role not emphasized in many analyses. That is one reason for employing the term "ecological power." To focus on ecological power is to highlight the role of the state in coordinating and allocating resources. The authority of the state may not be used *overtly* as an instrument of coercion, except on a limited and occasional basis. Instead, people look to the state as a mechanism for establishing and maintaining orderly arrangements. The authority of the state may serve mainly to designate the rule under which social transactions occur—when market processes are given free rein, when limited, and when displaced by some other form of allocation. Because transaction rules are too indirect in consequences, too remote and abstract in character, or too technical in form to elicit the mobilization of most groups, public officials have considerable leeway. They are not thrown into the role of being broker between groups in direct conflict, at least not at the level of making ecological adjustments in transaction rules.

If social complexity moves the democratic state away from the possibility of rule by a unified elite and also away from the role of group-conflict broker, then what role is played? I suggest that the state is an instrument for coordinating and allocating resources. Laissez-faire rhetoric, ideologically inflated arguments about the virtues of market allocation, and characterizations of the modern state as primarily a wielder of coercive control miss what is otherwise obvious: The state plays the major role in setting society's priorities. It does so, not by the enactment of a comprehensive plan (no such plans are enacted) and not even by the adoption of a budget (though budgets are part of the overall process), but rather by determining which rules of transaction apply. No single set of actions determines these rules, and no single level of government monopolizes the determination. Tax policy, direct and indirect subsidies, regulatory rule-making, reporting requirements, and informal cooperation with private activities combine to determine the conditions under which many vital social transactions occur. In a sense, modern government is about pushing and pulling levers—increasing tax incentives for one activity, adding tax disincentives to another, encouraging some forms of competition, discouraging others, setting in motion some bargaining processes, stultifying others, subsidizing some activities and withdrawing subsidies from others. These are not highly coordinated activities and they follow no comprehensive plan, but they structure and restructure the social ecology in which we live and operate.

Mutual adjustment plays a role, of course. Some groups push in one direction and others push back in the opposite direction.

But even here government has a say; it can determine the rules under which mutual adjustment occurs—collective bargaining is an obvious example. But the point in talking about social complexity is to emphasize the large degree to which social actions have significant indirect effects that are not amenable to treatment through *mutual* adjustments. And the point in talking about ecological power is to emphasize that adjustments come about on a basis other than spontaneous occurrence or individual responses to market or other processes. Power actors are at work trying to modify the terms under which various transactions take place. As they do so, they are pushing one set of considerations to the neglect of others. Thus, while direct bargaining between mutually affected bodies is no longer the principal device at work, power activity is still present. Effective participants in the ecological-power game form partnerships with government. Out of such partnerships they obtain considerations that impose few, or at least small, direct costs on others. The notion of social ecology instructs us, however, that indirect consequences are also important, especially cumulatively.

The old notions of power centered on control, resistance, and bargaining are inadequate to account for the emergence of these partnerships. The term "ecological power" guides us toward an examination of the conditions that favor some private interests over others for entry into partnership with government. Working deductively from the nature of social complexity and inductively from the urban experience, I have suggested that the possession of strategic information and the control of serviceable and dependable resources are the key elements in ecological power.

NEXT STEPS: DIRECTIONS FOR FUTURE RESEARCH

Traditionally, power has been discussed primarily in terms of the intentions and motivations of social actors under conditions of conflict. Command, resistance or countercommand, and resolution—these were the behaviors to be examined and accounted for. Even studies of non-decision-making tended to focus on why resistance or countercommands were missing from some scenarios of social action.

Particularly because non-decision-making is hard to operationalize, there is a tendency to regard many activities as other than power phenomena. After all, much of what happens in a complex society appears to be unintended, and in many respects it is. Individual actors are busy pursuing limited goals, and many are either unaware of the large consequences of their actions or perhaps unable to do more than protect themselves within given conditions. However, the argument here is that not everyone is helpless when it comes to modifying these social ecologies. Hence, we need to talk about ecological power.

Ecological power is parallel to simpler forms of power (particularly non-decision-making) in some ways but not in others. Ecological power is a competitive advantage in modifying conditions over which no one has complete control or full understanding. It is the capacity to make adjustments that are not mutually desired, though many of those affected are simply nonparticipants in the process of ecological adjustment.

What kind of research agenda grows out of this concept of power? A twofold effort is in order: a "forest" aspect and a "trees" aspect. Let me start with the trees. First, there is the matter of identifying the individual instances of ecological adjustment. In the area of community development policy, for example, we can determine what is left to the market, what is handled by nonmarket transactions, and what is handled by mixed criteria. Community development policy is, of course, only one area of policy, albeit an especially important one. School attendance districts and practices are another significant area of policy—one that is, in this case, fairly remote from the immediate concerns of business interests.

Market versus nonmarket transactions are also only one category of transaction. The degree to which planning and decision making are fragmented versus integrated on a metropolitan-area basis is another set of transaction rules especially worthy of consideration. In this case, comparisons can be made across policy areas—transportation versus education, for example.

It is crucial *not* to look at the social ecology only at a given time. The exercise of ecological power is a process of making continuing adjustments over time. It is important, then, to see which efforts are sustained and which are only sporadic and unsustained. And it is important to see which interests are involved in sustained efforts, which in sporadic efforts, and which are largely uninvolved. These categories of interest can then be compared to the forms of information they command and the kinds of resources they control. The notion of ecological power also invites greater attention to policy analysis and evaluation. Though it is no easy matter, it is worth investigating the broad social impacts of various kinds of ecological adjustment (see Long, forthcoming, on the related topic of social accounts). Bradley, Downs, and Small (1982), for example, have found that most urban transportation and economic development policies have no beneficial results for many city residents and may even leave some worse off.

Let me now turn to the forest. The term ecological power provides a way of thinking about structural advantage and disadvantage. In this sense, it follows a path already cut by various Marxists. Structural Marxists, of course, emphasize quite different components from what I have suggested here. They focus on the contradictions between system needs, such as those that bring about tension between capital accumulation and social legitimacy (see, for example, O'Connor, 1973). One obvious line of inquiry, then, would be to examine the relative merits of treating structural advantage as ecological power, as is done in this chapter, versus treating structural advantage in Marxian terms. Another appropriate line of inquiry would be to explore the political implications of structural advantage. If there are structural imperatives, what does that mean? For example, is community development policy simply the unfolding of laws of determinism? If not, what are the leverage points? How can advantages and disadvantages be altered? Here we are at a different level from that of how to alter a particular situation. We are asking whether relatively powerless interests can in some way be empowered to be participants in ecological power. If not, does this mean that power is simply the mechanistic working out of the rules of a given system? If so, does this mean that only revolutionary change is possible? If powerlessness can be remedied by lesser means, how? The ecological power argument directs attention to information, organization, and resource capacity. Are these elements that can be altered to bring about a more even balance of ecological power? Structural advantage is relative, and, compared to transient conditions, it is relatively stable. But we should not assume, without examination, that structural advantage is either absolute or fixed. Pursuing the possibilities of informed and intended change seems a worthy item on our research agenda. After all, at least as far back as Aristotle, creating the "good" political life has been a concern of political science. The notion of ecological power does not change that concern; it only increases our appreciation of how complicated that pursuit may be.

REFERENCES

ANTON, T. J. (1963) "Power, pluralism, and local politics." Administrative Science Quarterly 7 (March): 425-457.

BACHRACH, P. and M. S. BARATZ (1970) Power and Poverty. New York: Oxford University Press.

BANFIELD, E. C. (1961) Political Influence. New York: Free Press.

BENDIX, R. (1960) Max Weber: An Intellectual Portrait. Berkeley: University of California Press.

BRADLEY, K. L., A. DOWNS, and K. A. SMALL (1982) Urban Decline and the Future of American Cities. Washington, DC: Brookings.

BREWER, G. D. (1973) Politicians, Bureaucrats, and the Consultant. New York: Basic Books.

CARO, R. A. (1974) The Power Broker. New York: Knopf.

CLARK, P. B. (1969) "Civic leadership: the symbols of legitimacy," in O. P. Williams and C. Press (eds.), Democracy in Urban America (2nd ed.). Chicago: Rand McNally.

CRENSON, M. A. (1971) The Un-Politics of Air Pollution. Baltimore: Johns Hopkins University Press.

DAHL, R. A. (1982) Dilemmas of Pluralist Democracy: Anatomy vs. Control. New Haven, CT: Yale University Press.

DOMHOFF, G. W. (1983) Who Rules America Now? Englewood Cliffs, NJ: Prentice-Hall.

DOWNS, A. (1967) Inside Bureaucracy. Boston: Little, Brown.

DYE, T. R. (1983) Who's Running America?—The Reagan Years. Englewood Cliffs, NJ: Prentice-Hall.

EMERSON, R. M. (1962) "Power-dependence relations." American Sociological Review 27 (February): 31-41.

FAINSTEIN, S. S. et al. (1983) Restructuring the City: The Political Economy of Urban Development. New York: Longman.

FOSLER, R. S. and R. A. BERGER (1982) Public-Private Partnership in American Cities. Lexington, MA: D. C. Heath.

GAVENTA, J. (1980) Power and Powerlessness. Urbana: University of Illinois Press.

GOFFMAN, E. (1961) Asylums. Garden City, NY: Doubleday.

HARDIN, G. (1968) "The tragedy of the commons." Science 162 (December): 1243-1248.

HARDIN, R. (1982) Collective Action. Baltimore: Johns Hopkins University Press.

HENSON, M. D. and J. KING (1982) "The Atlanta public-private romance: an abrupt transformation," in R. S. Fosler and R. A. Berger (eds.), Public-Private Partnership in American Cities. Lexington, MA: D. C. Heath.

HUNTER, F. (1953) Community Power Structure. Chapel Hill: University of North Carolina Press.

JONES, B. D. and L. W. BACHELOR (1985) "Local policy discretion and the corporate surplus," in R. Bingham and J. P. Blair (eds.), Urban Economic Development. Beverly Hills, CA: Sage.

KELLER, S. (1963) Beyond the Ruling Class. New York: Random House.

LaPORTE, T. R. (1975) Organized Social Complexity. Princeton, NJ: Princeton University Press.

LIPSKY, M. (1970) Protest in City Politics. Chicago: Rand McNally.

LONG, N. (1958) "The local community as an ecology of games." American Journal of Sociology 64 (November): 251-261.

———(forthcoming) "Getting cities to keep books." Journal of Urban Affairs.

LOWI, T. J. (1979) The End of Liberalism (2nd ed.). New York: Norton.

LUKES, S. (1974) Power: A Radical View. London: Macmillan.

LYND, R. S. and H. M. LYND (1937) Middletown in Transition. New York: Harcourt Brace Jovanovich.

McCONNELL, G. (1966) Private Power and American Democracy. New York: Knopf.

O'CONNOR, J. (1973) The Fiscal Crisis of the State. New York: St. Martin's.

OLSON, M. (1971) The Logic of Collective Action. Cambridge, MA: Harvard University Press.

PETERSON, P. E. (1981) City Limits. Chicago: University of Chicago Press.

SCHATTSCHNEIDER, E. E. (1960) The Semi-Sovereign People. New York: Holt, Rinehart & Winston.

SCHELLING, T. C. (1971) "On the ecology of micromotives." Public Interest 25 (Fall): 61-98.

STONE, C. N. (1976) Economic Growth and Neighborhood Discontent. Chapel Hill: University of North Carolina Press.

———(1980) "Systemic power in community decision making." American Political Science Review 74 (December): 978-990.

———(1984) "New class or convergence?" Power and Elites 1 (Fall): 1-22.

———R. K. WHELAN, and W. J. MURIN (1986) Urban Policy and Politics in a Bureaucratic Age (2nd ed.). Englewood Cliffs, NJ: Prentice-Hall.

SYKES, G. M. (1958) The Society of Captives. Princeton, NJ: Princeton University Press.
TABB, W. K. and L. SAWERS (1984) Marxism and the Metropolis (2nd ed.). New York: Oxford University Press.
THOMPSON, J. D. (1967) Organizations in Action. New York: McGraw-Hill.
WEINBERG, M. W. (1977) Managing the State. Cambridge: MIT Press.
WHITT, J. A. (1982) Urban Elites and Mass Transportation. Princeton, NJ: Princeton University Press.
WILDAVSKY, A. (1979) Speaking Truth to Power. Boston: Little, Brown.
WRONG, D. H. (1980) Power: Its Forms, Bases, and Uses. New York: Harper Colophon.

PART III

The Pluralist View of Community Power

5

Community Power and Pluralist Theory

ROBERT J. WASTE

Pluralism remains the dominant paradigm used to explain the distribution of power in American society. The popularity or ideational hegemony of pluralism is based in part on the ability of pluralism to explain four different configurations of power in American society. The present chapter argues that the modern pluralist tradition contains four distinguishable schools of thought. The unique characteristics of each pluralist variant are analyzed and described. In addition, the four pluralist variants—classical pluralism, hyperpluralism, stratified pluralism, and privatized pluralism—are ranked along an inclusiveness dimension using public contestation as a ranking criterion. There are good reasons to believe that differences among the pluralist variants are significant, both for the intellectual history of ideas and for empirical research in community power and national policy studies.

STATEMENT OF THE PROBLEM

As Pitkin has demonstrated so ably in her studies of representation and justice, words are both claims about the world and

AUTHOR'S NOTE: I wish to thank William Domhoff, Thomas Dye, Dale Rogers Marshall, and Clarence Stone for their helpful suggestions on earlier drafts of this chapter.

tools for our use (Pitkin, 1967, 1972). The query at hand is: What claims are we making when we use the word "pluralism"? And when are we using the tool correctly and most effectively? If the community power literature is any indication, pluralism is—somewhat ironically—being used too narrowly *and* too broadly. On one hand, pluralism is used as a general purpose generic term to describe virtually all political situations not encompassed by two other generic terms—elitism or class (structural) analysis (Manley, 1983). As Manley (1983: 368) observes: "Pluralism, elitism and class analysis have divided students of power for decades, but there is little doubt that pluralism is the dominant theory or paradigm of power among American social scientists." In fact the widespread popularity of pluralism may be more a consequence of this broad definitional quality than of any descriptive superiority pluralism may possess when compared to the competing elitist or class analysis theories.

On the other hand, pluralism is frequently used to describe myriad idiosyncratic community and national power scenarios or approaches to the study of community power. One leading participant in the pluralist dialogue of the 1960s and 1970s has argued that by 1971, pluralism had "become so protean a word that it should be retired from serious discourse on grounds of excessive ambiguity" (Wolfinger, 1971: 1102). This plea for a "decent burial" notwithstanding, the classification of pluralist theorists and the creation of new pluralist variants or subtheories now represents a small cottage industry for urbanists in particular and social scientists in general (Polsby, 1980; Kariel, 1961, 1970; Connolly, 1969; Ricci, 1971, 1984; Greenstone, 1975; Stone, 1981). Useful distinctions have been drawn between traditional or "Golden Era" and "modern" pluralists (Garson, 1978; Jones, 1983; Ornstein and Elder, 1978; Greenstone, 1975), between types of modern pluralists (Kelso, 1978; Polsby, 1980; Manley, 1983), and between the traditional or modern pluralist mainstream and such recent pluralist variants as hyperpluralism (Wirt, 1974; Lineberry, 1980) or "street-fighting pluralism" (Yates, 1977).

Thus, in the increasingly pluralistic universe of pluralist discourse, there is no shortage of pluralist theories, subtheoretical spin-offs, or classificatory schema. Why, then, should we add to this confusion with yet another typology? And what would we hope to gain by such an attempt? Despite the widespread use of the pluralist paradigm to describe national, cross-national, and local political scenarios, there is little general agreement on what pluralism per se denotes. What claims is a speaker making when he or she labels country *A* or city *x* as pluralistic? And what usage of the term is correct? Put differently, when are we employing the *tool* pluralism correctly?

Pluralism has a basic set, or cluster, of claims (we refer to this set of claims from now on as "classical pluralism") that distinguishes it from other paradigmatic terms—notably, elitism and class analysis. Nonetheless, while pluralism, like any other tool, may be put to several uses, only four of these are most appropriate to national and community power studies and to social science generally. While other appropriate uses for the term pluralism will continue to emerge, social science usage currently falls into four recognizable camps. The purpose of the present chapter is to clear up some of the confusion surrounding pluralism by clarifying the basic set of claims associated with the term and distinguishing between this basic set and other subsets or variants that have also emerged as appropriate uses of the pluralist paradigm.

Focusing on pluralist descriptions of power at both the national and local levels, the present chapter distinguishes between the classical pluralism of Madison, Tocqueville, Bentley, Riesman, and Truman, and the more modern hyperpluralist, stratified, and privatized variants of such pluralist theorists as Calhoun (1851a, 1851b), Beer (1974), Wirt (1974), Dahl (1953, 1961, 1971, 1978, 1979, 1982), Dahl and Lindblom (1976), Lindblom (1977), Polsby (1960a, 1960b, 1969, 1972, 1979, 1980), Wolfinger (1960, 1971, 1974), Wildavsky (1964, 1984), McConnell (1966), and Lowi (1967, 1969; see Table 5.1). Aside from providing a lexicon of leading uses of the term "pluralism," the chapter constitutes a brief argument on the correct usage of pluralism in both community and national power studies. In brief, I would argue that the assertion "community *x* or country *A*

TABLE 5.1
A Typology of Pluralist Theorists

	Pluralist Variant			
	Classical Pluralism	*Hyperpluralism*	*Stratified Pluralism*	*Privatized Pluralism*
Theorists	Madison Bentley Riesman Truman (I)	Calhoun Beer Wirt Yates Lowi Lineberry	Truman (II) Dahl Lindblom Polsby Wildavsky Wolfinger	McConnell Bauer Dexter Pool
Key Concepts	*Groups:* Multiple	Strong	Activist	Few
	Government: Brokerage	Weak	Brokerage	Private

is (or is not) pluralist" should be followed by a second assertion or inquiry seeking to determine what type of pluralist community locale x (or A) is (or is not) held to be. The assertion that locale x is pluralist is less helpful than the more precise assertion that locale x is a classical pluralist, hyperpluralist, stratified pluralist, or privatized pluralist polity. These terms, used correctly, allow listeners to know more precisely which claims pluralist scholars are advancing and, again more precisely, which empirical tests future scholars might wish to employ to replicate or refute the pluralist claims.

CLASSICAL PLURALISM

Asserting that locale x is pluralistic in the generic sense of the term, pluralistic is not without meaning. Quite the reverse. Such a reference probably denotes that locale x possesses a cluster of qualities usually associated with the term pluralism. Second, such a remark constitutes a claim that other equally broad paradigmatic generic terms (such as elitism or class analysis) do not describe political conditions in locale x.

In general usage, the claim that locale *x* is pluralistic usually denotes that the polity in question may accurately be described as an example of classical pluralism. Thus pluralism used in its broadest sense as a generic term refers to the classical variant of the pluralist paradigm. Classical pluralism, associated with such writers as Madison, Tocqueville, Bentley, Riesman, and Truman, holds that public policy is a tug of war between various interest groups that often ends in a delicate balance or compromise. Classical pluralism focused on group activity, positing that: (1) group activity is the key to policy formation, (2) groups broker with each other to form public policy, (3) no group or coalition of groups can long dominate the policy arena because of the existence of (4) multiple (Madison), overlapping (Truman), or yet-to-be-formed "potential" groups (Bentley), or because unwritten norms or "rules of the game" (Truman) promote (5) compromise or equilibrium policy outcomes (Rothman, 1960; Olson, 1965).

In the classical pluralist variant, politics is an open and free process, characterized by a give and take and either the spirit or the necessity of compromise. Thus the view presented is not unlike a large billiard table in which the different, and more or less equally sized, billiard balls bounce off one another until they eventually lie at rest. In the classical pluralist view, this eventual equilibrium is the policy outcome. Furthermore, all or most community members are represented in the policy process either directly or indirectly.

The billiard table or classical pluralist variant has been subjected to serious criticism, questioning both its descriptive accuracy and normative desirability (Rothman, 1960; Kariel, 1961, 1970; Olson, 1965; Lowi, 1967, 1969; Connolly, 1969; Hale, 1969; Ricci, 1971, 1984; Garson, 1978; Kelso, 1978). The remainder of the present chapter attempts to contribute to this ongoing dialogue by distinguishing between classical pluralism and other pluralist variants prevalent in the literature of community power and national policy studies. Central to the present chapter is the assertion that the classical pluralist variant, in its role as a generic term, has dominated the pluralist landscape.

While the existence of a general purpose pluralist description of community politics is useful, it is also deceptive, in that the classical variant has served as a rallying point, or lightning rod, that obscures other pluralist variants of equal descriptive importance for community power studies. Frequently, pluralism is discussed in monolithic terms. In some studies the classical variant has come to emerge as pluralism per se or as pluralism "writ large." Thus pluralism (read: the classical pluralist variant) will be contrasted with elitism (Dye, 1984, 1985; Falkemark, 1982; Manley, 1983), criticized as a tradition with a unified (in some cases homogenized) body of writers (Rothman, 1960; Olson, 1965; Kariel, 1961, 1970; Connolly, 1969), or bluntly chopped into one or two camps, such as "Golden Era" and later (Garson, 1978), traditional and "modern" (Jones, 1983), or "Pluralism I" and "Pluralism II" (Manley, 1983).

Many of these distinctions are useful—especially those which distinguish between pluralist writers on a chronological basis—and yet the time or date of the various pluralist treatises is less helpful for distinguishing between various writers within the pluralist school than is a second dimension—inclusiveness. By inclusiveness, we refer to the level of participation by citizens in the policymaking process (Dahl, 1971: 6-16; McConnell, 1966: 365). Inclusiveness refers to the mobilized interest of community residents. Beyond mere interest (a position or information on the policy topic), how many citizens have mobilized that interest and achieved a role in the policymaking process? How many groups and individuals have succeeded in having some impact on the formation of policy *A* in polity *x*?

HYPERPLURALISM

If we were to rank the four major pluralist variants on a continuum ranging from broadly inclusive to narrowly inclusive, they would fall along the continuum in the order represented in Figure 5.1. Wirt (1974) has described a second pluralist variant in which city government is characterized, as in the first variant, by

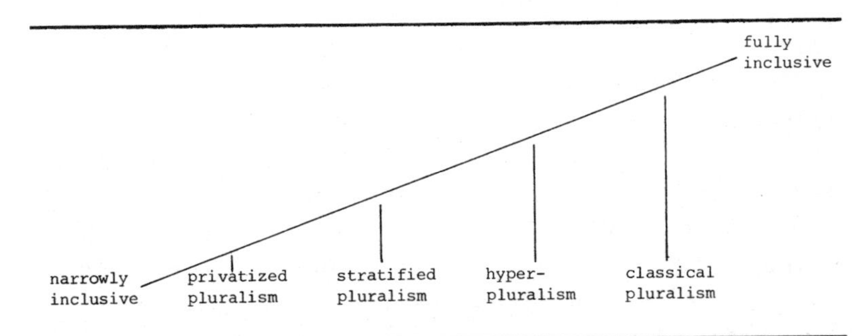

Figure 5.1 The Variants of Pluralist Theory Ranked on Inclusiveness Continuum

multiple groups but where the government itself is weaker than the pressure group system. Government in such cities lacks the "means or the will to resist any of the many competing demands that barrage it" (cited in Jones, 1983: 190). Wirt (1974: 350) cites Wood's classic study of 1400 governments in the New York-New Jersey metroplex in the late 1950s where local governments were "players at a roulette wheel, waiting to see what number will come up as a result of decisions beyond their direct control." Lineberry (1980: 50) has extended Wirt's hyperpluralist variant:

> Hyperpluralism, as the name implies, is an exaggerated, extreme, or perverted form of pluralism; it is so decentralized and pluralistic that it has trouble getting anything done. . . . Too many groups, each refusing to take "no" for an answer, are at the root of hyperpluralism.

Several scholars have described national and subnational politics in what we may now call hyperpluralist terms. These include Calhoun, Beer, Gardner, Lowi, and Yates. Writing in the *Disquisition on Government* (1851) and the *Discourse on the Constitution and Government of the United States* (1851), Calhoun envisioned a national government held in check—most importantly on the issue of slavery—by state and local interests, each exercising a "negative" or veto over national policy decisions.[1] This negative role, or check, by interest groups might play itself out in one of several ways. In one hyperpluralist scenario, Beer describes the pre-Thatcher interest group politics of Great

Britain as "stop and go politics" wherein one interest group captures the government or carries the day momentarily, only to be replaced by a succeeding interest group that reverses earlier policy (such as Britain's on-again, off-again pre-Thatcher monetary policies) or replaces the governing body with a party of a different ideological hue. Second, Beer uses the phrase "pluralistic stagnation" to describe a situation in which national or local policymakers are frozen or paralyzed into inaction. In such cases, the pressure group system is stronger than the government, leading to what John Gardner (1978) calls "a paralysis in national policymaking." In an earlier analysis, Lowi (1967) describes policy blockades, which we have described herein as hyperpluralist—in which group demands are strong and government policymakers are decentralized and weak—as "interest group liberalism."

A city characterized by hyperpluralist politics, then, would be a polity with strong groups, weak government, and either quick reversals in city policies or policy stagnation. Yates (1977: 33-37), representative of the hyperpluralist school, has argued that hyperpluralism—or, to use Yates's own term, "street-fighting pluralism"—has contributed to the problem of urban fragmentation which has made many modern U.S. cities "ungovernable." Consistent with other members of the hyperpluralist approach, Yates (1977: 34) argues that for many American cities, "there is no coherent administrative order to implement and control public policy. . . . [Instead, there is] a political free-for-all, a pattern of unstructured, multilateral conflict in which many different combatants fight continuously with one another in a very great number of permutations and combinations."

TRUMAN (II) AND THE TRANSITION TO STRATIFIED PLURALISM

Yates and the other hyperpluralists do not share one of the key assumptions of classical pluralists. For the classical theorists, including the early work of Truman, political groups were roughly equal. If they did not share equal resources, they could form alliances to bring, in Kenneth Galbraith's (1952) words, "power from across the market." As Truman admitted, groups varied in their "strategic access" to policymakers, but those lacking such access could form alliances with groups possessing entré. Failing that, low-resource groups were still protected, because if policymakers dangerously violated societal norms ("the rules of the game"), "potential groups" on the fringe of society would form to right the balance. This notion of an internal compass, satirized later by Galbraith (1973) as a reification of Adam Smith's "invisible hand," is rejected by the hyperpluralists, who note: "The fact that diverse, variable and complex interests infuse urban policy making does not mean that each interest has equal power or even substantial power. What is meant is that urban policy makers are faced with, and must react to, a diverse, variable, complex range of demands and that the various urban political actors, whatever their power, must fight it out as best they can" (Yates, 1977: 37).

In his introduction to the second edition of *The Governmental Process* (1971), Truman broke ranks with the classical pluralists. A dean at Columbia University in the turbulent 1960s, Truman abandoned the notion that the system produced roughly equal results for all groups. Potential groups could, and did, fail to mobilize to protect either the "rules of the game" or less powerful groups. Second, political society was stratified into two groups—the differentially weighted groups who did participate in American politics, and an underclass of "chronic nonparticipants." Thus pluralism did occur, but it did so in a rarefied arena in which the participants had various levels of "strategic access" to decision makers. Even flagrant violations (such as the denial of the civil rights of American blacks) of the interests of the politically unorganized might be tolerated by the pressure group players. These political outcomes would—if unchallenged by players in the pressure group system—remain unchallenged by a large group of persons who, instead of playing in the system and losing, simply did not play at all.

STRATIFIED PLURALISM

The break with classical pluralism by Truman in 1971 was anticipated in 1961 by Dahl and given systematic expression in the latter's study of power and politics in New Haven. In *Who Governs?* Dahl developed what we shall label "stratified pluralism" to describe a city which Dahl characterized somewhat curiously as "a republic of unequals." Dahl advisedly chose the phrase "republic" over "democracy." He thought New Haven to be a pluralist city, but as with Truman (II) and the hyper-pluralists, the pluralism involved players of different weights,[2] resources, and attitudes than the general or nonparticipating public. The cornerstone of Dahl's theory was the empirically demonstrable assertion that New Haven, and most American cities, are divided into two strata: the political actives (*homo politicus*) and the larger group of politically inactive persons (*homo civicus*). Thus, in New Haven, politics was pluralistic but stratified. New Haven involved a pluralism for and among the active citizens, as opposed to a free-wheeling pluralism involving the larger or general public as a whole.

Interestingly, the stratification in New Haven did not always follow predictable lines. Despite a generation of election studies which have shown a strong correlation between voting and high SES (Berelson, Lazarsfeld, and McPhee, 1954; Campbell, Gurin, and Miller, 1954; Campbell, Converse, Miller, and Stokes, 1964; Milbrath, 1965; Nie, Verba, and Petrocik, 1979) or middle to older age (Wolfinger and Rosenstone, 1980), Dahl's 1961 study showed an anomaly. Instead of a strata of *homo politici* comprising higher SES and older citizens, in at least one measure of participation (voting), blacks in New Haven had higher rates of participation in campaigns and elections than did whites. In all, 31% and 13%, respectively, of blacks had "high participation" or "highest participation," compared to 15% and 5%, respectively, for whites.

This relative overparticipation by blacks in the New Haven electoral process was unmatched by a similar high rate of participation in other aspects of political life (such as city schools or city redevelopment decisions). Echoing arguments by V. O. Key, Jr. (1949, 1955, 1966) presented at the same time as the *Who Governs?* study, and arguments raised later by Nie, Verba, and Petrocik (1979), Dahl argues that citizens move in or out of politically active—and thus pluralistic—strata due to perceived interests and structural opportunities. The New Haven blacks, for example, faced fewer impediments to electoral participation than to economic or social participation. Rather than concluding that the voting percentages indicated the wide acceptance of blacks as active and legitimate forces in the New Haven community, Dahl argues that black voters were compensating in their voting habits for inactivity and illegitimacy in other aspects of New Haven life. He notes: "Although discrimination is declining in the private socioeconomic sphere of life, New Haven Negroes still encounter far greater obstacles than the average white person. . . . In contrast to the situation the Negro faces in the private socioeconomic sphere, in local politics and government the barriers are comparatively slight" (Dahl, 1961: 293-294). Thus black voting patterns in New Haven and the urban North in general were examples of the use of "slack resources" by the black community. Having few other resources than their vote and their numbers, black community activists urged northern blacks to compensate for discrimination in other areas by concentrating their participation in voting, an area in which their impact was capable of producing results.[3]

Thus, not only was New Haven stratified into political actives and inactives, but the membership of the two strata changed across issues. In each case, the pluralism of New Haven involved pluralism among the actives, but the membership in the activist strata changed from issue to issue. This, then, is the key to stratified or activist pluralism. As opposed to the free-wheeling, more inclusive pluralism depicted by classical pluralist writers, Dahl describes a pluralism which is stratified and limited largely to the activist strata. American cities may be pluralistic, but this observation has to be tempered with the caveat that the majority of pluralist activity occurs within the political strata of *homo politicus*, and that the membership and representativeness of the political strata may change—both across issues within one city and across cities on the same issue. Hence, within stratified pluralist cities, these may be said to be more or less pluralistic. To phrase the proposition differently, city *x* may be more pluralistic than city *y* because the size and representativeness of the pluralistic political strata in *x* may be larger than the political strata on the same issue area in city *y*.[4]

PRIVATIZED PLURALISM

The three previous models of pluralism—classical, hyperpluralism, and stratified pluralism—represent descending points along a continuum of inclusiveness or community involvement and participation. Stratified pluralism, as noted earlier, acknowledges the existence of large numbers of nonparticipants in city politics, while the hyperpluralists and classical pluralists envisioned large numbers of citizens and groups participating in the daily politics of national and civic life. In *Private Power and American Democracy* (1966), McConnell sets forth a theory of "privatized" pluralism. The touchstone of privatized pluralism is the finding that in many policy arenas, a limited number of participants have usurped the authority and resources of public policymaking for private ends. In an examination of the role of agricultural, business, and labor groups in the national policy process, McConnell argues that all three groups had privatized to varying degrees national policymaking on issues of concern to them. For McConnell, all three groups: (1) exercised varying degrees of private power over public institutions and resources, (2) blurred the line between public and private (labor less so because of its stress dating back to Gompers on autonomy and voluntarism), and (3) attempted to force government to respond to the secular needs and the narrow constituencies of each group.

The picture presented by McConnell is that of pluralism come, temporarily, unglued. In a policy process generally pluralistic—in the classical, hyperpluralistic, or stratified sense—McConnell discusses the deviant case in which a small group manages to privatize public policymaking. Although this variant of pluralism would seem close to elite or class interpretations of power in American society, there are important differences. For class theorists, there is a class of "economic notables," defined by its ownership of the means of production, which exercises dominance in policy formation. For elitists, there is a circle of notables who stand atop bureaucracies and tend to come together out of common career experiences, common problems running large bureaucracies, and (for Mills and several other elitists) common social origins.

Still, a nuance should be emphasized—one that separates McConnell from other pluralists examined earlier. McConnell's view, which envisions the capturing of pieces of the government policymaking apparatus by different groups that are not challenged, makes him a different kind of pluralist than those who see government policy as the outcome of competing interest groups. Additionally, McConnell's nonelitist bent is evident in his claims that the various separate elites are not united, do not care about big or general issues, and do not control parties, the presidency, or the Supreme Court.

McConnell holds that business interests as a whole have not decentralized and privatized the national policy process in the entire arena of business policy. Rather, he agrees and cites Berle: "There is no high factor of unity when several hundred corporations in different lines of endeavor are involved. . . . Dominance by them is not a present possibility." In McConnell's words: "Instead of a conquest of government as a whole, control of significant parts of it may be established by particular business interests" (1966: 254). Although business is not monolithic in its influence, nor successful in decentralizing and overwhelming governmental decision arenas, it remains successful at the national level by institutionalizing "the ideal of industry collaboration with government" as a "long-standing tradition on which the Department [of Commerce] was founded" (McConnell, 1966: 271). McConnell also argues that business sectors have been successful in reducing the number of players in their specific policy arenas. Thus business policymaking may appear to be pluralistic in the Department of Commerce. It may, in fact, appear to be classically pluralistic, a "town meeting approach." But more important, this is pluralism for the few. Unlike classical pluralism, the size of the active constituency has been greatly reduced from the classical proportions by business design and to the advantage of business interests.

Thus, for McConnell, American national policymaking involves an array of interests and arenas, some of which have been privatized and others not. Some have been decentralized and overrun by parochial interests, while other national policy arenas have successfully resisted privatization despite a structural decentralization in their operations and decision processes. For McConnell, the point is that not all arenas have been privatized. Thus the importance of political parties, the presidency, the Supreme Court, and Congress, as well as state and local parties and institutions, is as vehicles to cause the arenas overrun by the "small constituency" to be more responsive to the large constituency, or the general public.

For McConnell, the prescription to resolve situations in which privatized pluralism has arisen is not decentralization but more pluralism. Concerned that decentralization facilitates policy privatization for such groups as business and agriculture, McConnell prescribes a stronger role for political parties. Serving as broad, catch-all intermediary groups, stronger political parties might resist the more parochial interests of producer and consumer groups, both nationally and at the local level. McConnell notes the paradoxical legacy of the Progressive era reforms which served to weaken parties and legislative bodies via such means as nonpartisan ballots and the legalization of initiative, referendum, and recall elections. These reforms have aided local and national policy interests, on one hand, to insulate their concerns from a political party that might force them to merge their concerns within the context of a larger political coalition. On the other hand, the initiative process has encouraged a new class of special interest lobbyists to circumvent the traditional hearing and legislative process while legislating secular concerns (such as Proposition 13) indirectly via the initiative ballot process.

Bauer, Pool, and Dexter, in *American Business and Public Policy* (1963), present a view of group activity consistent with the privatized pluralism of McConnell. In a study of trade and tariff policies in the Eisenhower and Kennedy administrations, Bauer et al. argue that business groups often have great influence in one area of policy that affects their business and revenues directly, but that even groups with established links ("issue networks," to use Heclo's term) to policy centers (1) did not have or seek comparable influence across the board in business or agricultural policymaking generally, and (2) could have their established position in their own narrow policy area disrupted by a change in national values or priorities (labeled by Bauer et al. the "ideology of the times"). The latter occurred when, as Lowi (1964: 699) notes:

> Congress and the "balance of power" seem[ed] to play the classic role attributed to them by pluralists. Beginning with the reciprocity in the 1930's, the tariff began to lose its capacity for infinite disaggregation because it slowly underwent redefinition, moving away from its purely domestic significance towards that of an instrument of international politics. . . . The significant feature here is not the international but the regulatory part of the redefinition. As the process of redefinition took place, a number of significant shifts in power relations took place as well, because it was no longer possible to deal with each dutiable item in isolation.

The findings of Bauer et al. constitute a corollary to the privatized pluralism of McConnell. Not only is privatization uneven, it is also dynamic and subject to change. Where privatization occurs, and despite its degree of present strength, privatized policy arenas may be forced to widen their "small constituencies" due to several factors, including changes in institutional relationships, types of actors present, or the prevailing "ideology of the times." The view of pluralism expressed by the privatized pluralists stands at the far end of the inclusiveness continuum. The actors in the policy arena are few in number; they are a small constituency. The policy may be constructed by bargaining and compromise, but bargaining and compromise among the few. Unlike the view of the world presented by most elitist and class analysts, the influence of key interest groups in society will be uneven for privatized pluralists.

For such pluralists, the demonstration that one interest group sits atop a bureaucracy with proven policy influence does not lend credence to a second conclusion—namely, that most policy bureaucracies are governed by small privatized constituencies. Indeed, the development of privatized pluralist theory suggests that such periods of influence and control are episodic. Writing of the New Deal era, McConnell argues that by the 1940s, agricultural interests had privatized the national policy process almost completely; in effect, preempting elected and appointed policymakers. Yet Bauer et al., writing in the 1960s, argue that the agricultural policy sector is populated by groups only marginally connected to each other in terms of policy alliance and policy influence. Even these groups might be upset in their own parochial policy area by changes in actors, institutions, or the ideology of the times.

For privatized pluralists such as McConnell and Bauer et al., privatization may occur in national or local policy arenas. It is not necessarily permanent, and not necessarily coordinated with or related in any meaningful way to activities of other peak associations in American society. This, then, may mark a cardinal difference in the world as viewed by the pluralists—even pluralists of the privatized stripe—and elitists or class analysts. For elitists and class analysts, coordination between members of one privatized arena and members of other privatized arenas in pursuit of common policy outcomes is probable. For pluralists, the relationship is, at most, problematic. Put differently, an arena which is privatized today may be hyperpluralist tomorrow. In each case, the question is a matter for empirical research.

CONCLUSION: THE FOUR VARIANTS OF MODERN PLURALISM

The pluralist legacy in America can be traced back to James Madison and Alexis de Tocqueville. In the longer view, pluralist elements existed in several theorists from the Italian Renaissance to the Scottish Enlightenment. Indeed, some analysts have traced pluralist leanings in Machiavelli, Locke, Montesquieu, Rousseau, and Adam Smith (MacIver, 1957: 144-148; Odegard, in Bentley, 1967: vii-xxviii). The present chapter has attempted to group American pluralist theorists along a descriptive continuum ranging from broadly to narrowly inclusive policymaking arenas. Rather than presenting pluralism as a homogeneous theory, chronologically developed, with each succeeding theorist staying within, if amending, an earlier tradition, the American pluralists have been grouped along the inclusiveness continuum. Thus pluralist writers have been described as falling within one of four modern pluralist camps: classical pluralism, hyperpluralism, stratified pluralism, or privatized pluralism (see Table 5.1). Pluralist (and class or elitist) analysts and community power researchers are urged to take note of the differences within the pluralist camp and to incorporate these differences into their own research and thinking. Thus it becomes necessary at times to distinguish between pluralism as a generic term (read: the classical pluralist variant) and various local or national pluralist scenarios (read: the classical, hyperpluralist, stratified, or privatized pluralist variants).

To reiterate a simple point made earlier, the assertion that a given city—or country—is pluralistic is a useful social science claim. Such comparisons at the level of broad generalities between elitist and pluralist policy scenarios are often helpful. Indeed, a given city or country may be accurately described as pluralist instead of elitist, or vice versa. However, such broad terms are less useful when claims are advanced not across categories (for instance, city x or country A is elitist or pluralist) but within categories (all cities x_1 to x_{10}, all countries A_1 to A_{10} are pluralist). If, for example, the assertion is made that several cities are pluralistic, the typology of pluralist variants discussed here will provide a basis for explaining which variant of pluralism (and hence which set of claims) is being associated with which cities. Thus an observer may disagree with this author's claim that most cities in the United States are pluralistic, but he or she will be able to disentangle the earlier claim from a second favorite claim of the author—namely, that the pluralist politics in Minneapolis and Portland is more classical than the hyperpluralism of San Francisco, which in turn is more inclusive than the stratified pluralism of San Diego or the privatized pluralism associated with the Windy City in the days when Richard Daley held sway.

NOTES

1. It is jarring but accurate to include Calhoun on a list of American pluralists. The inclusion of Calhoun in the pluralist camp is traditional (see Orenstein and Elder, 1978). The pluralism of Calhoun was, of course, an exclusionary affair. Calhoun, an apologist for racism, sought to broaden the base of American politics and increase the power of local elites by reviving and refining the doctrine of interposition raised by Jefferson in the Kentucky Resolution (1798) and Madison in the Virginia Resolution (1799). Calhoun sought to create a system in which local plantation-based elites could freeze or nullify the actions of a national policy arena alternately dominated by the national Democrats led by Jackson, the national Republicans, and the newly formed Whig party nominally led by Webster and Clay.

2. A point worth stressing is that—despite the impression given in the secondary literature—all pluralists hold that groups are differentially sized and weighted. The theoretical difference among pluralists involves a range of opinions from the more sanguine (classical pluralists) to less sanguine (privatized pluralists) about the ability of the political system to even out such differences and produce equitable results (see Appendix A in Dahl, 1982).

3. The picture of New Haven that Dahl presents is that of a city divided into two strata of actives and inactives. These strata involved the political, economic, and social worlds of New Havenites. Key to Dahl's depiction and central to understanding him as a pluralist is his assertion that the political stratum—when contrasted with the economic or social strata—was easier for "organized and legitimate" interest groups composed of the disadvantaged in New Haven to penetrate than were the social or economic stratum. Unquestionably "impediments" (Dahl, 1977) existed to voting and meaningful representation, but compared to economic issues such as racial covenants, the practice of redlining by New Haven banks, or black employment in general, the political stratum was infinitely more amenable to penetration by organized and active lower-SES groups.

4. One area of critical importance for stratified pluralists and pluralist theorists in general is the development of empirical measures to demonstrate the propositions that: (1) polity *x* is more pluralistic than polity *y*; and/or (2) polity *x* is more pluralistic on issue *A* than those on issue *B*.

REFERENCES

BAUER, R. A., I. de SOLA POOL, and L. A. DEXTER (1963) American Business and Public Policy. New York: Random House.

BEER, S. (1974) The British Political System. New York: Random House.

BENTLEY, A. F. (1967) The Process of Government. Cambridge: Harvard University Press. (Original work published in 1908)

BERELSON, B., P. F. LAZARSFELD, and W. McPHEE (1954) Voting. Chicago: University of Chicago Press.

CAMPBELL, A., G. GURIN, and W. E. MILLER (1954) The Voter Decides. Evanston, IL: Row, Peterson and Co.

CAMPBELL, A., P. E. CONVERSE, W. E. MILLER, and D. STOKES (1964) The American Voter. New York: John Wiley.

CONNOLLY, W. E. [ed.] (1969) The Bias of Pluralism. New York: Lieber-Atherton.

CRENSON, M. (1971) The Un-Politics of Air Pollution. Baltimore: Johns Hopkins University Press.

DAHL, R. A. (1961) Who Governs? New Haven: Yale University Press.

———(1971) Polyarchy. New Haven: Yale University Press.

———(1977) "On removing certain impediments to democracy." Political Science Quarterly 92 (Spring).

———(1978) "Pluralism revisited." Comparative Politics 10 (January): 191-204.

———(1979) "Who really rules?" Social Science Quarterly 60 (June): 144-151.

———(1982) Dilemmas of Pluralist Democracy: Autonomy vs. Control. New Haven: Yale University Press.

———and C. E. LINDBLOM (1976) Politics, Economics, and Welfare (2nd ed.). New York: Harper & Row. (Original work published 1953)

DOMHOFF, G. W. (1967) Who Rules America? Englewood Cliffs, NJ: Prentice-Hall.

———(1974) The Bohemian Grove and Other Retreats: A Study in Ruling Class Cohesiveness. New York: Harper & Row.

———(1978) Who Really Rules? Santa Monica, CA: Goodyear.

———(1983) Who Rules America Now: A View for the 80's. Englewood Cliffs, NJ: Prentice-Hall.

DYE, T. R. (1976) Who's Running America? The Carter Years. Englewood Cliffs, NJ: Prentice-Hall.

———(1984) Understanding Public Policy. Englewood Cliffs, NJ: Prentice-Hall.

———(1985) Politics in States and Communities. Englewood Cliffs, NJ: Prentice-Hall.

FALKEMARK, G. (1982) Power, Theory and Value. Lund, Sweden: Liber Gleerup.

GALBRAITH, J. K. (1952) American Capitalism: The Concept of Countervailing Power. Boston: Houghton Mifflin.

———(1967) The New Industrial State. Boston: Houghton Mifflin.

———(1973) Economics and the Public Purpose. Boston: Houghton Mifflin.

GARSON, G. D. (1978) Group Politics. Beverly Hills, CA: Sage.

GREENSTONE, J. D. (1975) "Group theories," in F. Greenstein et al. (eds.), The Handbook of Political Science, Vol. II. Reading, MA: Addison-Wesley.

———[ed.] (1984) Public Values and Private Power in American Politics. Chicago: University of Chicago Press.

HALE, M. Q. (1969) "The cosmology of Arthur F. Bentley," in W. E. Connolly (ed.), The Bias of Pluralism. New York: Lieber-Atherton.

HECLO, H. (1978) "Issue networks and the executive establishment, " in A. King (ed.), The New Political System. Washington, DC: American Enterprise Institute for Public Policy Research.

HERKERS, J. (1978) Article. New York Times (November 12): 1.

JONES, B. D. (1983) Governing Urban America: A Policy Focus. Boston: Little, Brown.

KARIEL, H. S. (1961) The Decline of American Pluralism. Stanford: Stanford University Press.

———(ed.) (1970) Frontiers of Democratic Theory. New York: Random House.

KELLER, S. (1979) Beyond the Ruling Class: Strategic Elites in Modern Society. New York: Random House.

KELSO, W. (1978) American Democratic Theory: Pluralism and Its Critics. Westport, CT: Greenwood Press.

KEY, V. O., Jr. (1949) Southern Politics in State and Nation. New York: Knopf.

———(1955) "A theory of critical elections." Journal of Politics 17 (February): 3-18.

———(1966) The Responsible Electorate. Cambridge: Harvard University Press.

LAZARSFELD, P. F., B. BERELSON, and H. GAUDET (1944) The People's Choice. New York: Duell, Sloan & Pearce.

LINDBLOM, C. E. (1977) Politics and Markets. New York: Basic Books.

LINEBERRY, R. L. (1980) Government in America: People, Politics, and Policy. Boston: Little, Brown.

LOWI, T. (1964) "American business, public policy, and political theory." World Politics 16: 677-715.

———(1967) "The public philosophy: interest-group liberalism." American Political Science Review. 61 (March): 5-24.

———(1969) The End of Liberalism: Ideology, Policy and the Crisis of Public Authority. New York: Norton.

MacIVER, R. F. (1957) "Interests," in E. R. Seligman (ed.), Encyclopedia of the Social Sciences, vol. VIII. New York: Macmillan.

MANLEY, J. F. (1983) "Neo-pluralism: a class analysis of pluralism I and pluralism II." American Political Science Review 77 (June): 368-383.

McCONNELL, G. (1966) Private Power and American Democracy. New York. Vintage.

MILBRATH, L. (1965) Political Participation. Chicago: Rand McNally.

MILLS, C. W. (1956) The Power Elite. New York: Oxford University Press.

———(1958) "The structure of power in American society." British Journal of Sociology 9 (March).

NIE, N. H., S. VERBA, and J. R. PETROCIK (1979) The Changing American Voter (2nd. ed.). Cambridge: Harvard University Press.

OLSON, M. (1965) The Logic of Collective Action. Cambridge: Harvard University Press.

ORNSTEIN, N. J. and S. ELDER (1978) Interest Groups. Lobbying and Policymaking. Washington, DC: Congressional Quarterly Press.

PITKIN, H. F. (1967) The Concept of Representation. Berkeley: University of California Press.

———(1972) Wittgenstein and Justice. Berkeley: University of California Press.

POLSBY, N. W. (1960a) "How to study community power: the pluralist alternative." Journal of Politics 22 (August): 474-484.

———(1960b) "Power in middletown: fact and value in community research." Canadian Journal of Economics and Social Science 26 (November): 592-603.

———(1969) "'Pluralism' in the study of community power: or erklärung before verklärung in wissenssoziologie." American Sociologist 4 (May): 118-122.

———(1972) "Community power meets air pollution." Contemporary Sociology 1 (March) 88-91.

———(1972) "Community power meets air pollution." Contemporary Sociology 1 (March): 88-91.

———(1979) "Empirical investigation of the mobilization of bias in community power research." Political Studies 27 (December): 527-541.

———(1980) Community Power and Political Theory (2nd rev. ed.). New Haven: Yale University Press. (Original work published 1963)

RICCI, D. M. (1971) Community Power and Democratic Theory. New York: Random House.

———(1984) The Tragedy of Political Science: Politics, Scholarship, and Democracy. New Haven: Yale University Press.

RIESMAN, D. (1967) The Lonely Crowd. New Haven: Yale University Press.

ROTHMAN, S. (1960) "Systematic political theory: observations on the group approach." American Political Science Review 54 (March): 15-33.

STONE, C. N. (1981) "Community power structure—a further look." Urban Affairs Quarterly 16 (June): 505-515.

TRUMAN, D. B. (1971) The Governmental Process (2nd ed.). New York: Knopf. (Original work published 1951)

WILDAVSKY, A. (1964) Leadership in a Small Town. Totowa, NJ: Bedminister Press.

———(1984) The Politics of the Budgetary Process (4th ed.). Boston: Little, Brown. (Original work published 1964)

WIRT, F. M. (1974) Power in the City: Decision Making in San Francisco. Berkeley: University of California Press.

WOLFINGER, R. E. (1960) "Reputation and reality in the study of community power." American Sociological Review 25 (October): 634-644.

———(1971) "Nondecisions and the study of local politics." American Political Science Review 65 (December): 1063-1104.

———(1974) The Politics of Progress. Englewood Cliffs, NJ: Prentice-Hall.

———and S. ROSENSTONE (1980) Who Votes? New Haven: Yale University Press.

YATES, D. (1977) The Ungovernable City: The Politics of Urban Problems and Policymaking. Cambridge: MIT Press.

6

From Labyrinths to Networks

Political Representation in Urban Settings

HEINZ EULAU

Social research is invariably contextual, reflexive, and latent. By "contextual," I am referring to the investigator's more or less self-conscious, implicit or explicit response to the field in which he or she works—its past and current theories, approaches, methods, and discoveries. This response may be accepting and extending, or it may be critical and rejecting. If the field is rent by intellectual controversies, the response may be one of more or less involvement in or avoidance of these controversies. Each of these modes of response is likely to have consequences that are difficult to foresee but that, in the end, define the place of a particular piece of research in the field.

By "reflexive," I mean the tendency of research, and especially of research carried on over many years by a team of investigators, to feed on itself. There is likely to be a tendency to both limit and extend the boundaries of the original research design in regard to theory, method, and substance as a result of experiences and reflections in the course of the research itself: It becomes evident that some things planned to be done cannot be done, while other things not planned can be done.

Finally, by "latent" I mean the intrusion in the course of the research of ideas or procedures either not overtly recognized or only vaguely sensed that, in retrospect, could have been used but were not, largely because the intuition involved is something emergent and hence, at the time, beyond the investigator's capacity to pursue. Unless one holds to the restrictive notion that social research should be a "closed system"—in other words, a process that moves from axioms and theorems to proofs and conclusions without deviation from the paths of "scientific logic"—all research is likely to have a latency component.

Our current research on social networks and political representation in an urban setting, the Redwood Network Project (RNP), begun in 1980 in cooperation with James H. Kuklinski of the University of Illinois, is largely a pilot or feasibility study. Because it grew out of our prior studies of representation, I shall briefly review the contextual, reflexive, and latent aspects of these earlier projects—more directly the City Council Research Project (CCRP) conducted in more than 80 cities of the San Francisco Bay Area between 1961 and 1972 (Eulau and Prewitt, 1973), and less directly the State Legislative Research Project (SLRP) conducted in four American states between 1955 and 1962 (Wahlke, Eulau, Buchanan, and Ferguson, 1963).

CONTEXTUAL ASPECTS

Let me emphasize at the outset that although CCRP's final publication (Eulau and Prewitt, 1973) was subtitled "Adaptations, Linkages, Representation, and Policies in Urban Politics," the project's original intellectual point of reference was not the urban field at all, and it was certainly not designed to address the controversies over urban power structures that, at the time of its incubation in the early 1960s, dominated urban research. The same is true of the current Redwood Network Project. Rather, the theoretical and empirical focus of attention, then as now, was the field of legislatures whose traditional institutional orientation was and is being complemented and supplemented by the newer behavioral approaches.

In fact, the original intention and purpose of CCRP was to replicate the largely individual-level role analyses of legislative and representational behavior pursued in SLRP. Like its predecessor, CCRP (and now RNP) hoped to make a contribution to generic knowledge about legislative behavior and processes, and the geographical location of the research was not decisive in the choice of city councils as sites or objects of the research, just as it had not been decisive previously in the choice of state legislatures as research sites. However, three lessons had been learned in SLRP that oriented the new project. First, if genuine comparative study of institutional structures was to be advanced, it was necessary to have as many units as possible available for analysis, and the units had to be easily (and economically) accessible to research. Second, the units had to be small enough to permit and facilitate observation not only of the behavior of legislators as individuals but also of their interactions, so that these interactions could be used as contextual variables in the analysis of individual behavior. Third, the units had to be analyzable as wholes on their own level of analysis, because their decisions are collective (rather than merely "aggregative") and their outputs are the products of the group rather than the group's individual members.

All of this is not to say that my CCRP collaborators and I were unaware of the work on urban or community politics by those interested in the shape and substance of local power structures and the controversies between elite, stratification, and pluralist theorists. However, at the very beginning of CCRP we made the purposive decision not to get involved in these controversies, interesting as they seemed to be. It seemed to us that the various sides in the debate over community power were captives of their models and the methodological constraints imposed on research by these models. To put it bluntly, we held the rather critical view that the different antagonists were out to find what they wanted to find and, indeed, found it. But we also seemed to have scientific warrant for our position. Whatever side of the community power issue particular investigators were on, it seemed to us that they all had to make asymmetric assumptions about the behavioral or structural patterns of politics that we were not willing to make. As our primary interest was not in power but in democratic representation, with its normative assumption of symmetry in the relationships between rulers and ruled, we had no use for the concept of power as a point of departure in our own work. We realized, of course, that in reality there are numerous constraints that limit any "playing out" of symmetry in representational relationships, and that because of built-in status differentiation, asymmetries will invariably occur. But we could treat this as an empirical question and not a theoretical one.

More influential than the community power literature in defining CCRP's research context was a tantalizing article by Lawrence J.R. Herson (1957) entitled "The Lost World of Municipal Government." Reviewing the literature (not including, of course, the community power studies which, except for Hunter, 1953, had not yet appeared), Herson had found that most texts on local government were antiquated, resting on theoretical and empirical foundations that were no longer valid. The study of city government, he pointed out, "lacking, for example, its first comparative study of the American city council—has yet to amass much of this necessary knowledge" (Herson, 1957: 340). Herson's comment on city councils thus reassured us that the city council as a research site had not been preempted and was open to investigation.

We were also much impressed by Janowitz's notion that local communities are what he called associations of "limited liability" (1952: 222). In this model the formal, public, and legitimate authority is an active, if not necessarily the most effective, role-taker in decisions on communitywide issues. In some respects this focus on a formal institution seemed rather old-fashioned, but as we proposed to study a great many cities and their councils, the assumption of the community as one of limited liability permitted us to remain open-minded in regard to the council's relative role as a decision maker. We could thus treat as moot the elitist assumption that the community's formal decision-making apparatus is one-sidedly manipulated by an often invisible and possibly conspiratorial power elite that makes the formal authority a mere pawn of the dominant interests; and we could avoid the pluralist assumption that diverse private interests can get together (openly or secretly) with other private interests on particular issues, make "arrangements" among themselves by way of bargaining or compromise, and have the formal authority simply referee their bargains or compromises. Rather, we assumed that under varying conditions the city council is a more or less active and effective participant in the games of community politics. Thus our assumption was that even if the council was only a minimal power agent in community decisions, not much could be done without it.

REFLEXIVE ASPECTS

Choosing city councils as research sites in order to study legislators' interactions and to use these interactions as explanatory variables in the analysis of individual behavior was part of CCRP's original research design. What was not part of the design was our final treatment of the council itself as a corporate actor or "object unit" at its own level of analysis. The prevailing notion in the behavioral study of politics at the time, it should be recalled, was that only individuals are fit to be studied by the methods of social science. This methodological individualism was reinforced by (1) the constraints of survey research—one cannot interview a group as a whole but only its individual members; and (2) the normatively individualistic bias of rational or purposive behavior theory. CCRP's transformation from a study of individuals into a study of groups at the macro level was thus the most significant aspect of the project's reflexivity.

We had assumed, of course, that intracouncil interactions among council members were unlikely to be random but would give rise to informal structures of various sorts, notably such typical decision structures as those later identified as unipolar, bipolar, and multipolar. But we also found, in a first-step analysis of individual interactions, that other dimensions of interpersonal relations were themselves structurally related, as well as related to the council decision structures. This led us to treating the council as a variable, ignoring for this purpose the individuals composing it. Our own initial methodological bias notwithstanding, we moved the entire analysis to the level of the council as a unitary or corporate actor in the several political processes that link the council to its environment as well as to its constituencies and clienteles, including the electorate.

What led to this strategy was, of course, the comparative focus of the research. We were sensitive to the fact that much knowledge was rapidly accumulating about the behavior of individuals *in* institutions but not about the behavior *of* institutions, or of other macro units of action (see Eulau, 1977). In the previous SLRP, some hesitant steps had been taken toward macro analysis. But in general, scholars like Almond and Verba (1963, in their five-nation study of civic cultures) relied largely on inferences from individual-level data to make statements about the behavior of macro units, possibly committing what has come to be called the "individualistic fallacy" (Scheuch, 1969). Of course, the "aggregation" or "composition" of the parts (the individual councilpersons, their attributes, attitudes, cognitions, and so on) into wholes that would facilitate comparative analysis of many institutional units in the macro perspective was not only dictated by methodological considerations; it also seemed justified by theoretical reconsiderations stemming from and being directed back to the project.

As legislative scholars, we were particularly interested in representation on one hand, and in the policies coming out of the representational process on the other hand. In the course of examining the individual-level data, it became apparent that while representation may certainly be an individual-level phenomenon involving relations between each councilperson and particular constituents, it is also a collective phenomenon: The legislature is not just an aggregate of individual representatives but also "a representative body," exactly as it is described in the traditional institutional language of political science, without being able, however, to demonstrate this holistic conception empirically. At this critical point in rethinking the project, we were aided enormously by the appearance of Pitkin's (1967) linguistic and philosophical study of the concept of representation. After a most thorough and minute explication of the concept's heterogeneous uses, Pitkin concludes that political representation is (1) best thought about as a collective phenomenon but (2) a phenomenon that, at the institutional level, might or might not emerge. On one hand, although a legislature is called a representative body, this does not necessarily mean that collective representation will occur. On the other hand, even if particular individual councilpersons are not responsive to their constituents, the council as a corporate actor can be. Turning these considerations reflexively back on the individual-level data, we decided on this theoretical ground to conduct the final analysis at the level of the group as a whole. (This is not the place to review the technical procedures involved, but see Eulau, 1971; Eulau and Prewitt, 1973: 30-60.)

Similarly, there was the recognition that public policies, emerging out of processes with "outside" individuals or groups, are genuinely integral properties of the legislative body. Once decisions are made and policies enacted, it no longer matters, at the macro level of analysis, how the policies came about, whether by unanimous or close votes, as a result of group or elite pressures, through managerial recommendations, and so on. At this point in our reflections on the data at hand, contemporary experiences in the arena of state politics once more intruded. A series of policy studies making use of budget data (notably Dye, 1966; Hofferbert, 1966; Sharkansky, 1969) had come to the incredible conclusion (see Salisbury, 1968) that politics had little to do with policy outputs—that most of the variance was explained by economic rather than political variables. The political variables used, however, were almost all of a macro-structural sort—party competition in the state, party control of the legislature, the relationship between the governor and the legislature, and so on. Missing in these studies, it seemed to us, was what we called councilpersons' "policy maps"—on the assumption that policies are made by people and not by structures, though the structures are likely to be constraints on what people can or will (and cannot or will not) do.

It seemed to us, therefore, that to explain and understand policies—in our case, revealed at the macro level by budgetary allocations over ten years for amenities and planning, where discretion is maximal; and for services, where discretion is minimal—we should also compose the individual-level data we had on councilpersons' policy perspectives into group-level data. Indicators of policies like budget allocations and expenditures are necessarily aggregate measures that in no way reflect the policy views of individual councilpersons, but only of the council as a group. Once we had identified what we called "council problem perceptions," "council policy positions," and "council policy images," we found that these composed variables could be meaningfully related to the aggregate measures of policy.

It also seemed to us at the time (see Eulau and Eyestone, 1968, where the analysis was conducted at the macro level, as against Eyestone and Eulau, 1968, where it was at the micro level) that the policy studies of the period suffered from two flaws: First, they were unduly "itemistic" in the sense that the attempt was made to relate particular budget items or categories to particular economic variables; second, they were unduly mechanical in assuming a linear relationship between policies and whatever structural political variables were employed to account for policies. Given our emerging emphasis on the council as a purposive corporate actor with an identifiable policy perspective or "policy map," we came to assume that councils are not only predisposed to respond to contemporary political contingencies but also to a developmental "policy environment" as indicated by the ebb and flow (that is, a not necessarily incremental process) of expenditures over time. The aggregation of such developmental sequences and their construction into a typology of policy environments involves a selective loss of microscopic detail. What justifies this loss is, of course, that this transformation of detail into a macroscopic configuration maximizes the explanatory power of policymaking behavior at the group level of analysis.

Finally, in the course of looking at the original individual-level data obtained through our surveys, it became apparent that it is extremely difficult to link the individual person, the council member, to such remote (that is, highly aggregated) variables as over-time electoral competition or to urbanization as a process through time. This would have required a multilevel type of analysis for which we were not prepared (the relevant statistical procedures were really not developed until the 1970s and are still in the process of development). This kind of linkage, it seemed to us, would be more visible and explicable if the council rather than the individual council member served as the object unit of analysis. While we did not use multilevel statistical procedures, we nevertheless assumed that the council is located in the middle of a hierarchy of levels and that unit properties from different levels could be transformed into properties at the group level of the council—that is, at its own level, where we could then employ standard correlation or regression techniques. At a lower level are the individual council members whose interactions constitute the group and make for group structures and functions of various sorts. At a higher level are the "environmental forces" of a given aggregate character—electoral competition, state of urbanization, or policies in some stage of development—that can be imputed to the council as properties of the group without committing the "ecological fallacy." Put differently, the council as a corporate actor was not only characterizable by its own (or integral or global) properties that are not reducible to the individual level, but also by properties of a relational, structural, or distributive sort constructed from individual-level properties, as well as by properties of an environmental kind that could be treated as contextual variables.

The point of all this is that sociopolitical research, if conducted over many years by a team of investigators, can be highly reflexive. Needless to say, perhaps, CCRP also produced a series of studies by its collaborators which were largely individual-level analyses (notably Black, 1970, 1972; Cronin, 1970; Eyestone, 1971; Loveridge, 1971; Lupsha, 1967; Prewitt, 1970; Zisk, 1973). The final product of the project (Eulau and Prewitt, 1973) by no means exhausted all the analyses that could have been conducted at the macro level, largely because, after many years, the investigators were exhausted and ready to go on to other things.

LATENT ASPECTS

These "other things" referred, in my own case, to some of the latent aspects of both SLRP and CCRP. Perhaps most latent in both projects was the great complexity of the individual and collective linkages that characterize legislative systems. Retrospectively, this complexity seems to exceed by far what either elite-centered (pluralistic) theories or empirical studies had led us to expect. We were therefore quite unprepared to cope with this complexity, although there were, in both projects, "points of departure" (or what is much better expressed in German as *Ansatzpunkte*, which literally translates into "points of beginning").

Just when I became aware of the potential utility of the concept of a social network as a tool with which to master social complexity is difficult to say. Rudiments of the notion can be found in *The Legislative System* (Wahlke et al., 1962), where we used the concept in the title of a chapter, "The Network of Legislative Role Orientations." But the use of the concept there was more metaphorical than indicative of what we actually did in identifying configurations of roles (rather than networks of individual legislators) or in describing and trying to explain corresponding "role structures." In fact, the linkages among legislators and between them and their constituents or pressure groups that we identified were all mediated by legislators' own multiple roles—that is, the analysis remained grounded in individualistic assumptions and explanations, even though the effort was made to close the macro-micro gap.

Rather curiously, the term "network" did not appear in *Labyrinths of Democracy* (Eulau and Prewitt, 1973), in spite of the fact that possible and actual lines of inquiry clearly pointed to its potential utility. Instead, in its search for linkages between rulers and ruled through the examination of what we called "constitutive" and "petitioning" processes, the notion of the city as a network of individuals or groups remained latent. This failure to make the idea of the city as a network manifest became clear to me shortly after the manuscript had gone to press. The occasion was an invitation to reflect on the relevance of *The Federalist* (Beloff, 1948) for a research design that would adumbrate representational linkages among individuals and groups on the three major levels—local, state, and national—of the American polity (Eulau, 1973). It then occurred to me that "linkage" is a term that is essentially binary and does not cover the possibility of "linkages among linkages" as I graphed it (see Figure 6.1), employing the term "network" in the title of the diagram. The current Redwood Network Project, briefly described in this chapter, is essentially the outcome of what, at the time, was at best a conceptual framework (although, for logistic reasons, RNP does not include the state level in its research design).

In some respects, our use of the concept of "labyrinth" in the epilogue to *Labyrinths of Democracy* symptomizes the latency of the network notion. Quite frankly, although we stated that we found "the metaphor of governance as a labyrinth preferable to other metaphors used in politics" (Eulau and Prewitt, 1973: 611), we might just as well have used "network," at least in its metaphorical sense. As we put it:

> A labyrinth is an enclosure with many entrances and exits. Its layout consists of a maze of pathways, but the pathways can be marked by signs that help one avoid false moves and lead the seeker where he wants to go. If the paths are not marked, trial and error may yet lead to discovery and the long way back. Democratic governance resembles the labyrinth. The labyrinth has walls that serve as boundaries but are more or less porous; it has major arteries and places of assembly but also byways, detours, nooks, and crannies. Passing through the labyrinth may take more or less time, depending on the continuities and discontinuities in the journey occasioned by what is known or unknown about the terrain [1973: 611-612].

This rich and colorful imagery of the labyrinth metaphor concealed our failure to make use of the scientifically warranted notion of network. I shall give just one example of how we could have translated a latent aspect of the analysis into a manifest one in network terms. Perhaps one of our most important findings

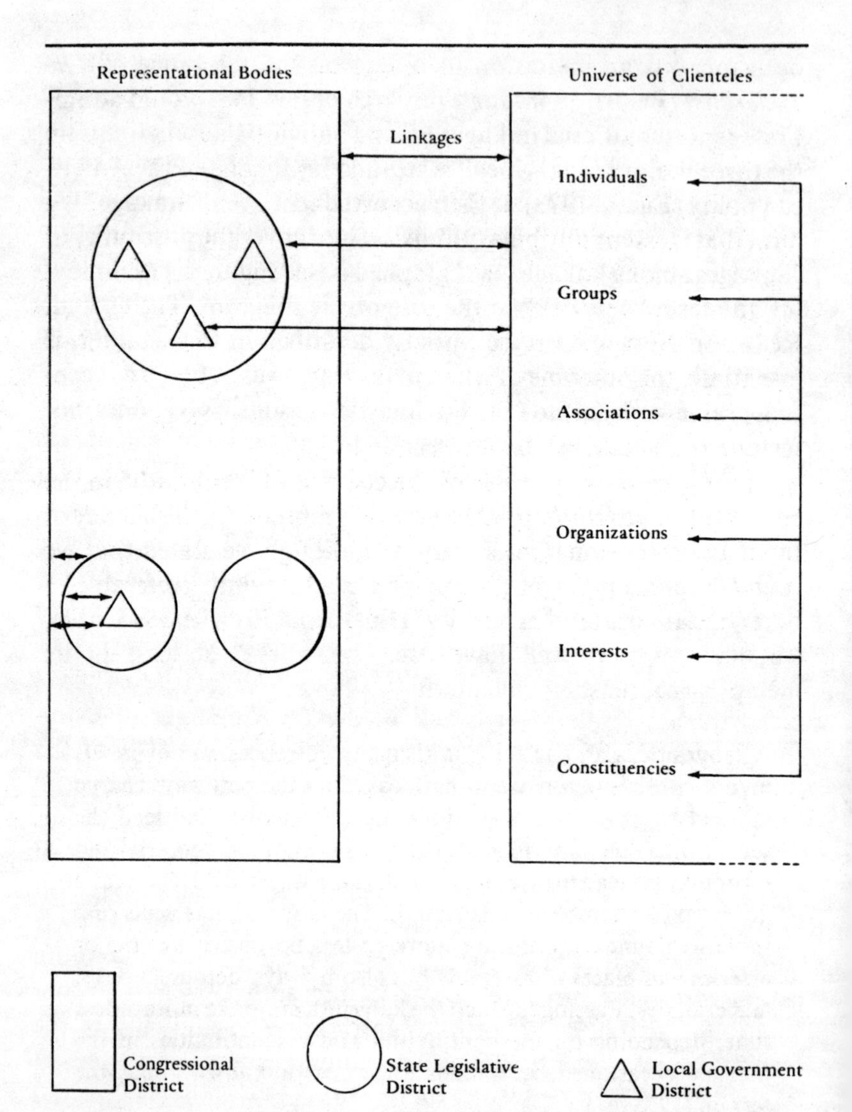

Figure 6.1 Network of Governmental and Clientele Linkages in Federal Representational System

was that a city's "organizational life begets organizational life" (Eulau and Pewitt, 1973: 316). The finding was significant not only in its own right but also because it seemed to falsify alternative hypotheses. For instance, one might expect that the activity levels of different social class-based organizations are negatively related, which would have meant that, as we put it, "certain types of groups preempt the organizational space of the community" (1973: 316). In fact, Stone (1980: 982), who, like us, assumes that the local community is socially stratified, has argued that what he calls the "systemic features" implicit in stratification "affect the power positions of competing groups with a stake in community decision making." He also argues that "opportunity costs" in influencing decisions are greater for the poorer than the wealthier groups, and so on.

Surveying our 82 cities, we found no support for this proposition, at least as far as the perceptions of their councilpersons provided the data (that of course could be biased, but the bias should counter rather than reinforce our findings). For instance, where the activity level of the Chamber of Commerce or the Jaycees—presumably higher-status groups—was high, the activity level of reform or protest groups and unions was also high (r = .74 to .86). When overall wealth of the community is taken into account, in "smaller" or "poorer" cities 67% of the councils reported moderate-to-high activity by the Chamber of Commerce, while in "smaller" and "richer" cities only 30% of the councils did so. On the other hand, in the larger cities overall wealth seemed to have almost no effect at all (79% of the councils reported moderate-to-high activity levels for reform or protest groups, as against only 18% in the larger and richer cities). But there was also some evidence of cumulative inequalities and their effect on group activity. For instance, in the wealthier cities more of the councils reported moderate-to-high activity by homeowner and neighborhood groups (60% in the smaller and 68% in the larger cities), while in the poorer communities considerably fewer councils indicated activity by such groups (33% in the smaller and 42% in the larger cities).

We concluded that in general, "greater activity by one type of group does not reduce the activity levels of other types; on the contrary, different types of groups tend to increase their activity levels supplementarily" (Eulau and Prewitt, 1973: 321). Rather than a city's overall wealth, correlational analysis indicated that city size appears to influence group activity patterns. Two reversals notwithstanding, we found the activity levels of various groups, whether of relatively high or low status, "to be much less interdependent in the simple small-town environment than in the more complex big-city environment" (1973: 323).

I have presented these findings and our interpretation, not in order to argue against assumptions concerning the impact of social stratification on political influence, but to suggest that aggregate correlational analysis concerning the interdependence of groups in various environments obviously makes latent assumptions about the functioning of social networks. If groups beget each other, it is not because they inspect correlation coefficients but because they perceive each other and, quite likely, interact with each other. Drawing on conflict theory, I would even argue that the more social and political conflict there is between groups occupying different ranks on a city's social status scale, the more likely it is that they are "bonded" by the conflict arising out of class interests.

The notion of urban democracy as a labyrinth was, of course, an afterthought. It was an implied admission that while we had made a variety of forays into the complexity of urban governance and policymaking, there had been no closure, no return to an easily identifiable starting point. What we were missing was the kind of red safety line that Alpine skiers let trail behind them, permitting rescuers to find them, dead or alive, if they should be caught by an unexpected avalanche. And the metaphor of a labyrinth was certainly not an appropriate model for tracing and connecting all of the components of the urban complexity that we had confronted. In the end, it seemed that we had presented some more or less connected chambers of the labyrinth but by no means all chambers and all connections. We could only claim that we had explored some important chambers—the councils as decision-making structures, representation as a function of "constitutive" and "petitioning" processes, city policy development as a confluence of both purposive collective action and environmental constraints, and so on.

On the CCRP's conclusion, it seemed to me that the attempt to understand urban governance was well served by making representation the centerpiece of research. But any attempt to come to grips with what is the basic problem of representation—the relationship between governors and governed, the few and the many—would require a more sophisticated and appropriate methodology, one that could bridge the gap between the traditional individualistic (or reductionist) and the more recent collectivist (or holistic) approaches to the puzzle of representation. This puzzle is essentially that by all measures now available and according to much anecdotal information at the micro level of observation, the linkage between the general citizenry, politically involved elites, and urban government appears to be weak in several respects: Contacts between governors and governed is periodic rather than continuous; attention by citizens to governance is limited rather than pervasive; responsibility of the governors is diffuse rather than specific; and so on. Nevertheless, at the macro level, representative government in American cities appears to be strong. But one must ask whether this puzzle is genuine and not illusory, more a matter of appearance than reality. Can it be that it is an artifact of inappropriate conceptualization or inadequate measurement?

In particular, is there a methodological-theoretical equivalent of the "red safety line" that would allow pursuing in even more systematic detail than was the case in CCRP the relational, structural, and behavioral processes that link governors and the governed? And if there is, would the methodology allow bridging the continuing gap between unduly individualistic interpretations of urban governance and policymaking on the one hand, and unduly holistic interpretations on the other hand? Network analysis in the study of urban representation recommended itself as a relational-structural approach to behaviorally and organizationally complex social phenomena that can "cover" in one analytic framework both individual and collective units of action and facilitate smooth transitions along a single continuum from the micro to the macro level of analysis—either by "moving up" the micro-macro continuum in constructing the whole out of its parts or "moving down" the continuum by reducing the whole to its various parts. In the remainder of this chapter, I hint at some of the theoretical underpinnings of representation conceived as a network phenomenon and briefly describe the research design of the Redwood Network Project.

REPRESENTATION AS A NETWORK PHENOMENON[1]

The methodological problem that the Redwood Network Project addresses is, as already indicated, the gap that continues to exist between micro and macro analyses of political representation. Micro studies, even if cast in aggregate statistical format, are predicated on individualistic, dyadic assumptions about the nature of the representational relationship (see Miller and Stokes, 1963, for the *locus classicus* of studies from the perspective of the represented; and Eulau et al., 1959, for studies from the standpoint of the representative). Macro studies, in turn, are predicated on holistic, collectivist assumptions theoretically articulated by Pitkin (1967) and operationalized by Prewitt and Eulau (1969; see also Weissberg, 1978).

Practitioners of either approach do not address the problem of relating micro-level interactions to macro-level patterns of representation. At the micro level, in particular, much has been learned about individual behavior, of representatives and represented alike; at the macro level, at least some insight has been gained into how the representational system is structured and functions, as well as why, despite weak links at the micro level, representative government is strong. But how the interactions between and among individuals, both representatives and represented, "compose" (rather than aggregate) to form "the over-all structure and functioning of the system, the patterns emerging from the multiple activities of many people" (Pitkin, 1967: 221-222) has so far eluded investigation and analysis. Treating representation as a network structure may be well suited to bridge the micro-macro gap by showing how and why social networks serve to transform the interactions and relationships among individuals into "the complex ways of large-scale social arrangements" (Pitkin, 1967:

221). These arrangements in turn are likely to facilitate or hinder what individuals do or do not and what they can or cannot do.

The theoretical underpinnings of our approach to social networks and representation will be stated simply. The most significant factor in the social environment of any person is another person. In fact, it is this existential condition that makes the individual a *persona*, a social actor who orients himself or herself to action by an awareness that other actors constitute important resources with which to obtain some purpose or goal. All other things being equal, the number of other persons whom an actor can call or count on as resources for achieving his or her goal, the status of these persons, and the quality or intensity of interpersonal bonds all define the social context in which a person acquires a perspective on the potential of politics and conducts himself or herself accordingly.

In this interactional perspective, political representation can be conceived as a process of resource mobilization for the attainment of goals—whether they involve "public-good" or self-interested policies or public or personal benefits. Because representation as a process also produces a relationship of some duration between the represented and representing persons, it tends to become institutionalized, with corresponding expectations among the actors involved. On one hand, the represented constitute a resource whose support the representative seeks to mobilize and maximize, and the representative is a resource that the represented seek to reach, influence, control, derive benefits from, and so on.

Our concern in this project is with a citizen's ability to contact another person as a resource when seeking to do something about some problem of importance to him or her. This "doing something" may, but need not, require access to and the intervention of a representative or representative body—in this study, the more or less typical city council or its individual members. To discover whether people see their political representatives or other persons as resources to achieve a goal is a major aspect of our research.

Thus representation need not involve a conscious relationship. A person may not have representation in view when trying to achieve a goal, whether of personal or public interest. A not altogether astounding finding of our preliminary inspection of the data is that relatively few people spontaneously see representatives as immediately available resources to meet some public problem of personal concern to them. But this does not mean that they are necessarily resource-poor as far as access to representatives or a representative body is concerned. Whether they are resource-poor or not, we argue, depends on the network of social relations in which they are involved (of which, of course, we can recover only a very small slice, with the slice defined by our research design). The critical point, in this theoretical perspective, is the citizen's ability or inability to contact others, regardless of whether they are representatives or not, to whom he or she can turn for consultation, advice, support, and so on in confronting a public problem that appears bothersome. Although the person so identified may not be a representative, he or she in turn may be connected, directly or also indirectly, with a representative.

Another assumption of the network theory of representation to be formulated is, of course, that networks not only link citizens to their representatives but also have some impact on their political roles, behavior, attitudes, and perceptions, just as it is an assumption of traditional "group theory" (Truman, 1951) that primary and secondary groups have some impact on their members. Insofar as group theory has been employed in order to understand and possibly explain representational behavior, relationships, and processes, analysts have viewed the group and its spokespersons as intermediaries between represented citizens and their particular representatives and representative body. In doing so, however, they made the further (but usually not explicit) assumption that, as the concept "group" implies, the individual is strongly tied to the group to which he or she belongs. They deduced from this assumption that the more primary or face-to-face the relations among group members are, the more effective the group will be as an intermediary in its ability to speak or act for its members vis-à-vis other groups, including government.

There is little empirical evidence in support of this hypothesis, and its truth value is an open question (see Olson, 1965).

The network theory of representation that we are groping for makes the opposite assumption. It assumes, and indeed postulates, that citizens enmeshed in close social relationships—that is, persons strongly tied to each other so that one may speak of "closed" social circles—are likely to be cut off from the representational process (unless, of course, all or many of them are highly and directly involved in politics, but as political involvement in the population as a whole is very low, the proportion of tight-knit and politically highly active networks is likely to be small). One may assume, therefore, that persons with weak ties or in loosely knit circles are more likely to reach the representative body than persons with strong ties in tightly knit networks. This is an adaptation to a representation of the network theory formulated by Granovetter (1973) and known as "the strength of weak ties" hypothesis.

Although the concept of a network is not unfamiliar to political scientists and has crept into the titles of a number of research articles (e.g., Crenson, 1978; Huckfeldt, 1983), it is mostly used as a metaphor to make sense of aggregate demographic data used to describe social contexts. All contextual analysis using aggregate data makes sense only if interactions between people can be assumed to occur. But as far as we know, actual interactions have not been traced at the level of individuals, at least not in survey research on the mass citizenry. There are a few studies that make use of respondents' own reports of the people they interact with in political matters, either in terms of categoric role types like relative, friend, or neighbor (see Eulau, 1980) or particular named persons (see Eulau and Siegel, 1981; Eulau and Rothenberg, 1986); but these studies do not then interview those named by respondents and are wholly dependent for network analysis on respondents' self-reports of their interactions with others.

More germane in regard to the problem of tracing linkages between representatives and represented are some observations made by Fenno (1978) in connection with his study of the "home style" of members of Congress. Fenno does not use the concept of network, but much of what he observed his subjects to be doing at home involved some sort of networking. He reports:

> One thing, however, is clear. All our observations and, hence, all our generalizations about congressman-constituent relationships involve a set of constituents far smaller than the total number in the district. No matter how a congressman allocates, presents, or explains, he reaches a relatively few people directly. Offsetting this situation, House members believe that as a result of their direct contacts with as many supportive constituents as they can reach, they will reach a great many more people indirectly. They are strong believers in the two-step flow of communication. They have to be. But they also think it works. Their belief is that though they may not reach as many people as they would like, those they do reach—whether easy or difficult—will talk to others about them [Fenno, 1978: 237].

Fenno quotes a congressperson who holds neighborhood coffee hours: "We send out seventy-five invitations. Only fifteen come, but word gets around that we've been there. There's a *ripple effect* in every community" (emphasis added). A ripple effect is what network analysis of representation is all about and, in fact, seeks to trace. Ideally, one would want to follow the indirect path of a representative's message as it moves directly from the source to someone with whom he or she is in direct contact, from that person directly to another person, and so on, until the chain breaks or closes. This was an aspect, at least partly, of our original research design that, for logistic reasons, had to be given up. On the other hand, by beginning at the citizen end of the representational chain, we may be able to reach at least some representatives. If city council member X is also a member of what we call citizen A's "action set," and A in turn belongs to citizen B's action set although representative X does not, there is an indirect, two-step link between B and X. If B, in turn, is directly linked to C, who is not directly linked to A and X, there is a more remote, three-step link between C and representative X. And so on (see Figure 6.2).

There is some risk in approaching the network analysis of representation from only one end of the representational rela-

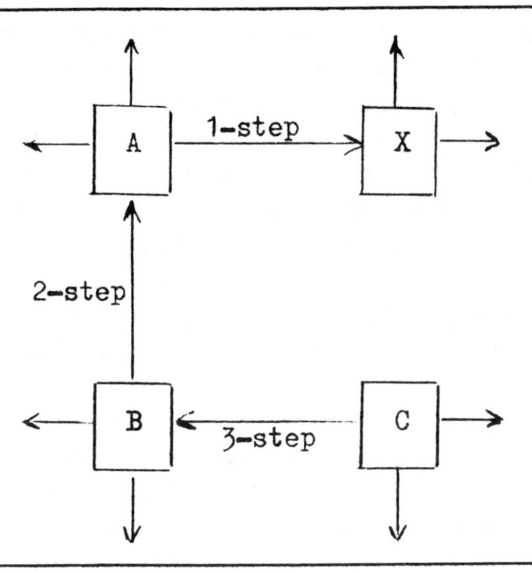

Figure 6.2 One-, Two-, and Three-Step Linkage Model

tionship—in this project, from the end of the represented—especially if the search for links is limited to a network root's "inner" (or first) and "outer" (second) zones (see Glossary of the network terms we are using). By concentrating on the represented citizen alone, we probably underestimate the ties that bind (if mostly indirectly) representative and represented. After all, "ordinary" citizens do not generally seek out and cultivate ties with their representative as the representative seeks out ties with constituents (which, however, may involve overestimation of the representational linkage structure). Moreover, the number of direct and indirect links one can discover is invariably constrained by one's research design. Had we extended the snowball interview technique into the network's outer zone (for which we had names given by persons in the root's action set), more links (stretching into a third zone) with representatives would probably have been reported. Our citizen-as-root networking procedure might thus make it appear that representatives are less tied to their constituency than they may actually be.[2] In any case, we encountered networks that do not seem to be directly *or* indirectly linked to the city council. This may be as much an artifact of the research design and the technique used to ascertain links between representative and represented as it may be a real-world condition, and we must necessarily treat it thus.

Although the focus of RNP is on representation as a social network phenomenon, the choice of an urban area as a "laboratory" for exploring networking as a research procedure in survey research on representation is not arbitrary. Even if the contemporary city is no longer a "community" in the halcyon sense of the word and more—as Janowitz put it more than thirty years ago, a community of "limited liability"—it is nevertheless "clearly not one of completely bureaucratized and impersonalized attachments. In varying degrees, the local community resident has a current psychological and social investment in his local community" (Janowitz, 1952: 223). Moreover, as Janowitz points out, "the large-scale organization of the urban community hardly eliminates the necessity of theoretical analysis of the networks of intimate and cohesive social relations which supply the basis of collective action on the local community level" (Janowitz, 1952: 222). How the citizen is linked to the local community's governance—the issue of representation—is therefore as significant for an understanding of civic life in the urban arena as the urban context as a setting for possibly "intimate and cohesive social relations" is relevant to an understanding of representation in a democratic society.

Design and Procedures

The Redwood Network Project was designed to explore the feasibility of network analysis at two levels of the American federal system of representation—the local and the national.[3] Without going into all the other aspects of the project not immediately relevant here, suffice it to say that what gave the project direction was more a quest for discovery than a quest for explanation (Diesing, 1972). Nevertheless, it was crucial for the project to proceed in a way that could identify social networks (which we assumed to exist) without violating those random sampling criteria that must be met if ultimately any kind of generalization about the part played by networks in the representational process is to be made. The difficulty in randomly sampling networks in the mass population is, of course, that unlike the case of individuals, membership groups, organizations, and institutions, the universe of social networks is unknown and, indeed, probably unknowable, possibly reaching infinity because networks appear anew or cease to exist almost continuously. There is no practical way of drawing a random sample of networks from a population of networks, because the boundaries of that population are unknown in advance (on snowball sampling, see Goodman, 1961; Granovetter, 1976).

What came to our aid in selecting a sample is the notion of "egocentric networks" or "action sets" which can be found in the anthropological literature.[4] Literally interpreted, an egocentric network or action set is composed of those persons in ego's social environment to whom he or she is directly tied in whatever social role and whatever substance the relationship may have. So interpreted, the concept of network largely remains a convenient metaphor (though it may be quite meaningful to ego), for it is yet to be known whether the persons tied to ego are themselves directly related to each other and not just indirectly through ego. Moreover, even if the persons in ego's action set—the term we prefer because it does not denote any linkage other than between ego and those with whom he or she has direct ties—are not themselves related, there is still the possibility that other persons are available as resources to ego by virtue of their own direct contacts outside ego's action set. That is, ego is indirectly linked through his or her own action set to persons in an outer or second zone. To limit analysis to action sets alone as units would greatly reduce the value that the network notion presumably yields in the study of a complex social process like political representation. A social network can therefore be defined as a set of action sets that are more or less connected to each other through the ties that exist between the persons in ego's action set and other persons in those ego-partners' action sets who are not in ego's first zone.

Most survey research seeking to discover social networks in the mass citizenry (and there is not much of it) has been limited to random sampling of individuals without follow-up snowball sampling. Each randomly picked person was asked to name others with whom he or she was in contact in some social role (relative, friend, neighbor, coworker, and so on) or in regard to some social function (from baby-sitting to voting behavior). The data so obtained gave strictly cognitive information about those named by the respondent—how close or frequent contact is, in what settings the contact occurs, what beliefs they hold, and so on (see Lauman, 1973; Sheingold, 1973). This kind of data yield action sets, but as the respondents are all randomly sampled, the probability of there being overlaps in personnel from one action set to another is rather remote. Moreover, respondents are usually asked to give only first names, either to assure anonymity or simply facilitate the interview process. The information thus given can be useful in characterizing the sampled individuals in terms of their self-reported social relations, but the method does not allow the tracing of connections between action sets.

The gamble in which we are involved lies in our application of network analysis in an arena whose population parameters are not known and whose boundaries are not given a priori. RNP also differs from those network studies of an empirical sort that are undertaken in arenas whose boundaries can be more or less precisely defined in advance. The "easiest" case, of course, is a small group (like a committee), all of whose members are known to know each other. At a more complex level, network analysis has been employed in the study of interaction in the professions, whose "members" can be easily enough identified by simply looking them up in a directory, even though no assumption can be made that all of the members actually know each other (see Coleman et al., 1966; Laumann and Heinz, 1984). Even more complexity has to be mastered by network analysis in the study of community elites whose identification is not necessarily given and whose links it is the task of network analysis to discover and graph for whatever purpose a study may be designed (see Laumann and Pappi, 1976). But even in this case the boundaries of the networks are ultimately specifiable, depending on one's definition of "elite" and the persons who, because of their personal attributes, are to be included.[5]

By way of contrast, in the case of networks in the general citizenry or mass electorate, the only feasible method is to randomly select individuals from among the citizenry who can serve as the starting points for snowball sampling. In a snowball sample the initial respondent is asked for the names of the persons he or she knows in some connection (defined by the purpose of the study) who can then be interviewed and asked the same questions as the root. Each of these snowball respondents has his or her own action set whose members may or may not be in the root's action set. The method makes it possible to observe three categories of linkages: first, between the root and his or her direct partners (the root's action set that also defines the network's "first zone" or, as we call it, "inner zone"); second, between the partners in the root's action set—the linkage that makes possible the empirical construction of the network's inner zone; and third, between persons in the root partner's own action set who are not in the inner zone (and were therefore not interviewed) but who are in what, from the standpoint of the root, is a second zone (which we call the "outer zone"). Some of the persons in the outer zone may be connected by virtue of having been named by inner zone respondents. They are referred to as "bridges."

For the purpose of RNP, we selected three census tracts in Redwood which, on the basis of demographic data from the U.S. Census, could be characterized as lower, middle, and upper class. Within each of these tracts we randomly drew a number of households—33 in the lower-class tract, 31 in the middle-class tract, and 29 in the upper-class tract, providing also for replacements to cover interview refusals by initially selected root respondents. Altogether, 131 persons were interviewed in the lower-class tract, 160 in the middle-class tract, and 141 in the upper-class tract. The administration of this kind of effort and the difficulties encountered have been described elsewhere (Eulau, 1984). The size of the action sets varied a great deal, as did the networks constructed from them.

Instrumentation

Underlying the instrumentation developed to discover a person's action set and to construct his or her social network was the assumption that representation is not something that exists but rather something that emerges when political actors, be they representative or represented, confront a situation that calls for some sort of action. As Pitkin (1967: 224) puts it, "Representation may emerge from a political system in which many individuals, both voters and legislators, are pursuing quite other goals." Representation conceived as an emergent process is predicated, of course, on the availability of governmental and political mechanisms that make emergence possible and permit relevant action. As Pitkin (1967: 233) remarks, "We can conceive of the people 'acting through' the government even if most of the time they are unaware of what it is doing, so long as we feel that they could initiate action if they so desired." In the case of city councils (and other legislative bodies in the American system), we can assume that the mechanisms are available for people to "initiate action if they so desired."

In order to discover some of the respondents' action sets—the other persons to whom they might turn as resources in order to initiate action in response to some "problem"—we confronted them with four hypothetical situations. We chose four situations because we assumed that people may turn to different persons for different purposes—that is, we assumed that representation is function-specific and, therefore, resource-specific as well. Somewhere in the middle of the interview, respondents were asked to whom they would turn for advice or help when confronting four "situations"—crime on the street, local industrial development, book censorship in the local library, and an unfair income tax law "passed by the national government in Washington." Later the respondents were read the names they had given and asked: "Can you think of any other people you might talk to or work with?" These names, if forthcoming, were added to the list. After all names had been recorded, the following question was asked:

> We would like to ask a few questions about each of the people on this list. Looking on page 13 of your booklet, could you tell me *how you know* (FIRST NAME ON LIST)—is he/she a relative, good friend, neighbor, work associate, someone you see at church, someone you see at school or community affairs, someone who belongs to the same group or organization as you, or someone in government?

The question was then repeated, in abbreviated form, of course, for each name on the list. Finally, two other questions were asked about each person on the list:

> Would you say that you know (FIRST NAME ON LIST) well, or not? Do you ever talk about politics and public affairs with (FIRST NAME ON LIST)?

Later in the interview the respondents were asked to "assist" while the interviewer filled in the addresses and phone numbers of each person on the list.

Some comment is desirable about the ordering of the network questions. We first asked the respondents to give (presumably) issue-relevant names without specifically directing them to the *kinds* of people they might turn to for discussion or action. We asked about the role relationship between the respondent and the persons in his or her issue-specific action set only after names had been solicited. The respondent was thus free to choose people from a variety of social contexts. This procedure differs from most similar inquiries into networks which are usually role-specific first. These inquiries also assume that a person may have several action sets, but they specify the roles in advance. Hence, if the question is about friends, it yields "friendship networks," and so on. Our primary interest is not in the role structure of a person's action set as such, although we also analyze action sets and networks in role-analytic terms. Rather, our objective is to discover whom, in his social environment, the respondent would perceive as a helpful resource in dealing with some issue of concern. In other words, our main focus is on those persons in the respondent's action sets who, in connection with particular policy issues or political problems, are "most likely to be sources of a variety of rewarding interactions," for these people "will be particularly important in shaping respondents' attitudes and behavior" (McCallister and Fischer, 1982: 78). By asking the issue domain questions first, we probably obtained more salient action sets than by asking first about categories of role-specific relationships. For instance, had we first asked respondents whether and to whom they might turn "in government," we would undoubtedly have received more names of government officials, including council members, than our question ordering elicited, but the salience of these names would have been in doubt.

Constructing the Networks

In the final study, the basic empirical units of network analysis are the individual persons who were interviewed. The units on which the construction of networks is based, however, are the particular *action sets*—that is, the sets of persons named by the respondents as "significant others" in the four issue domains, regardless of whether they were interviewed or not. The networks are constructed, therefore, not of individuals as building blocks but of the relations between and among them that emerge in the space created by the intersection of their action sets.[6] This two-dimensional space may be more or less "empty" or "full," depending on the individuals in the various action sets who "meet," so to speak, in the space. Figure 6.3 presents three hypothetical cases in the format of a sociometric matrix. Case A shows a space that is empty: There are six action sets—the root's and the five members of the root's action set, but there is no "network" because none of the respondents named either the root or each other, though they named other persons in the "outer zone" (from the perspective of the root). Case C represents a space that shows a great deal of interaction among the individuals in the root's action set, with some nominations being unilateral and others reciprocal. Case B is of a middling sort. Figure 6.4 presents an actual sociogram for RNP case #57 (which, by the way, probably falsifies the strength of the weak ties hypothesis).

Case A

	R	A	B	C	D	E	Z
R	-	1	1	1	1	1	-
A	0	-	0	0	0	0	6
B	0	0	-	0	0	0	3
C	0	0	0	-	0	0	8
D	0	0	0	0	-	0	5
E	0	0	0	0	0	-	7

Case B

	R	A	B	C	D	E	Z
R	-	1	1	1	1	1	-
A	0	-	1	0	0	0	8
B	0	0	-	1	0	0	5
C	1	0	0	-	0	0	2
D	0	1	0	0	-	0	9
E	0	0	1	0	0	-	3

Case C

	R	A	B	C	D	E	Z
R	-	1	1	1	1	1	-
A	1	-	1	0	1	0	6
B	0	1	-	1	0	1	9
C	0	1	1	-	1	0	3
D	0	1	0	1	-	1	5
E	1	0	1	1	0	-	8

Figure 6.3 Hypothetical Examples of Network Sociograms

	Interviewed							Not Interviewed					
	Root	101	103	104	108	109	110	102	105	106	107	AS SIZE	OZ SIZE
Root	--	X̲	X	X̲	X	X	X	X	X	C	X	10	
101	x̲	-					x̲					13	11
103			-							x		28	27
104	x̲			-		X						19	17
108					-							4	4
109			x			-						8	7
110		x̲					-			x		13	11
												95	77

Network Average Action Set Size = 13.57
Network Average Outer Zone Size = 12.83
Network Inner Zone Density = .1330
Network Inner Zone Centrality = .1614
Network Inner Zone Reciprocity = .1429
Network Inner Zone Hierarchy = .2000

Figure 6.4 Sociogram for Network 57

As Figure 6.4 shows, of the ten persons in Root's action set, six could be interviewed. The space created by the intersection of action sets shows three reciprocated (x underlined) nominations and four unilateral nominations, two of which were given to persons in the root's action set who were interviewed, and two to one person (#106) who could not be interviewed (and who is the community's federal congressman). In interpreting the various structural measures of networks, we do not of course base the interpretation on some absolute score but on the entire distribution of scores across all networks in our sample. As the two right-hand columns in Figure 6.4 indicate, in the case of this network the root's action set size is ten, but he or she is indirectly linked by one step to another 77 persons (that is, persons in the action sets of his or her own action set who constitute the network's outer zone).

Figure 6.5 transforms the sociogram of Figure 6.4 into a sociograph that more idiographically depicts the structure of network #57. It shows that the root names the city mayor (#103 and a member of the council) and the member of congress (#106 who could not be interviewed) as nodes in his or her action set. The mayor (#103) in turn named 27 persons, including six council members and the congressman, but not the root. The mayor (#103) was also named by respondent #109, while the congressman was also named by #101, who stands in a reciprocal relationship to the root as well as to #110. Both respondents #101 and #110 are directly linked to the council, with the former naming three council members, the latter one. Respondent #109, in addition to naming the mayor, also named two other council members. In other words, the root was directly linked to the council through the mayor and indirectly through three persons in his or her action set. Overall, then, network #57 appears to be relatively tightly knit but also relatively closely tied to the Redwood council. Of course, whether these preliminary nominal designations will be retained in the case of network #57 depends, as I suggested, on the sociometric profiles for all networks in the sample.

Space limitations do not permit me to describe in detail the measures we are developing for the structural properties of networks. They all derive from the location of individuals in the action sets found directly or indirectly in the space that we call the network's inner zone. Theoretical network analysis has formu-

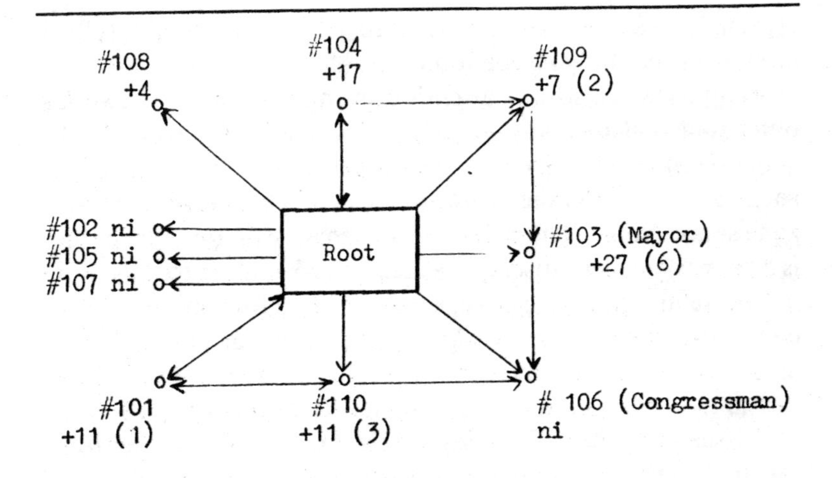

Legend

+ number of persons in network outer zone

( ) number of councilpersons among outer zone names

ni not interviewed

——→o unilateral nominations

o←——→o reciprocal nominations

Figure 6.5 Sociograph for Network 57

lated all kinds of such properties, such as density, reciprocity, centrality, and hierarchy. Density is a measure of the extent to which the network's inner zone is filled. Centrality emphasizes the overall similarity of the patterns that link the network's inner zone members. At the individual level, it is a measure of a person's "value" as a social resource and at the macro level a measure of the network's resource potential. Reciprocity is a straightforward measure of the ratio of actual reciprocal nominations in the inner zone to all possible reciprocal nominations. It is a partial indicator of the network's "tightness." Hierarchy is a very crude indicator of the network's status system in that it isolates what, in sociometric language, is called the network's "star." Other structural properties are the network's connectivity (of relevance especially to analysis of communication) and "clusterability" (of interest in the study of coalition formation).

Analysis will also pay attention to linkages in the network's outer zone that are made possible by "bridges"—for example, the situation in which A names X and B names X, but X is not in the root's action set and therefore not located in the network's inner-zone space. Somewhat different from these structural properties is a property called "discrimination." It refers more to the nature of the relations in a network and is therefore more akin to compositional properties like degree of acquaintanceship or amount of political conversation. It refers to the variety of names rather than their number given by respondents in regard to the four issue or problem domains. For instance, individual network members' attitudes on liberalism—conservatism, "social issues" and "economic issues,"[7] local industrial development and book censorship,[8] or party identification will serve this purpose. These measures at the group level are either averages or, more appropriately, their coefficients of variability.[9]

In addition to the "network questions," the RNP survey instrument also included a rich portfolio of questions about representation, group memberships and activities, and so on that I cannot describe here. To conclude, I present in Table 6.1 some stylized results of a preliminary analysis of a purposive subsample of 13 RNP networks in terms of just one network property—what we call its "knit" (density of relations among members).

If the profiles of tight- and loosely knit networks that emerge from the highly generalized and stylized configuration presented in Table 6.1 hold up in the final analysis, we would expect the networks to be significant variables in contextual analysis of citizens' representational behavior at the individual level. Tight- and loosely knit networks seem to expose the citizen to quite different social-structural conditions as he or she interacts with others. We seem to be observing that persons in tightly knit networks are quite differently constrained in their representational behavior, attitudes, and perceptions than persons in loosely knit networks, and that these constraints are internally coherent, all apparently stemming from the ways in which

TABLE 6.1
Configuration of Variables for Relationship Between Networks and Representation

Representation	*Property of Networks* Tightly Knit	*Loosely Knit*
Attitudes toward representation in general	Favorable: representation benign; trust, discretion given	Unfavorable: representation suspect; mandate preferable
Action potential of represented vis-a-vis representatives	Represented have posibility to influence decisions	Represented do not have choice but accept decisions
Distinction between public and private life and government intervention	No distinctions: private affairs involved with government	Distinction made: private affairs have little to do with government
Government vs. nongovernment help for problems in life	Nongovernment help not seen available; must rely on government	Help outside government seen available; need not rely on government
Norm of participation at cost of private interests	Rejected	Accepted
Local issue involvement vs. representation	No involvement; rely on representatives	Involvement anticipated; do not rely on representatives
City Council performance	More unfavorable assessment	More favorable assessment

networks are related to the external world. While loosely knit networks "radiate out" to the external world, tightly knit networks turn inward. This does not mean that tightly knit networks are any less permeated by the general political-normative culture. In fact, they seem to be more so precisely because, unlike the loosely knit networks, they have less of a possibility to "correct" for the culture.

As Table 6.1 indicates, persons in tightly knit networks seem to be more disposed to trust their representatives and let them act in their interest than persons in loosely knit networks. But they also seem to have a more efficacious view of their own ability to influence the representational decision-making process than persons in loosely knit networks. As a result, they see their own interests involved with (or mistake them for) public interests, and they see government as a benign patron on whom they can count. Citizens in loosely knit networks, having a more suspicious view of representation in general and feeling that their influence is limited, do not view government as benign and as an agent of their interests. As they see nongovernmental aid available, they do not seem compelled to rely on government as much as persons in tightly knit networks do. Trusting government and relying on government to advance their interests, persons in tightly knit networks reject the norm of participation and in fact expect not to get involved in local issues which they would rather entrust to the representatives. Persons in loosely knit networks behave in opposite ways. Paradoxically, citizens in tightly knit networks have a less favorable view of the city council as the ongoing local representative institution than citizens in loosely knit networks.

The paradox may be more apparent than real. We speculate that persons in loosely knit networks, more in actual contact with the council and seeing it doing a reasonably good job, need not entertain "idealistic" and stereotypic views of representation in general. They seem to know the limits of what the council can or cannot do for them. On the other hand, citizens in tightly knit networks, having exaggerated views of what representation can do for them and what they can do in the process but not really doing much about these views, see the council in an unfavorable light precisely because it does not conform to their preconceptions. As we said, we speculate in regard to all this, but we conclude from the exercise that even with our minute demonstration sample we have captured some aspects of political reality.

CONCLUSION

Because in the Redwood Network Project the approach to urban representation is from the observational standpoint of the represented citizenry by way of network analysis rather than, as in the City Council Research Project, from the standpoint of the representatives by way of compositional analysis, I hope that the former will overcome some of the difficulties encountered in the latter. Representational relationships between and among groups are necessarily mediated by individuals. If the potential of network thinking was not exploited in *Labyrinths of Democracy*, it was probably due to the fact that we had "lost" the individual person as we shifted analysis to the macro level of the council and the city. I believe, therefore, that some kind of multilevel network analysis is the more appropriate strategy to cope with the complexity of urban political representation. And I come to the apparently (but only "apparently") paradoxical conclusion that the more complex a macro unit seems to be, the more necessary it is to investigate its parts and their interactions at the micro level of analysis. This is where empirical (not just metaphorical) network theory and analysis come in.

GLOSSARY

Root: The person selected by way of random sampling as the starting point of a snowball sampling and interview process whose purpose is to identify a social network.

Action Set: The set of persons in an individual's social environment to whom he or she can turn in order to achieve some objective. The persons in an individual's action set may or may not be themselves linked.

Link or Tie: The connection between two or more persons. A link or tie may be direct (one step) or indirect (two or more steps; see Figure 6.2).

Inner Zone: The space (matrix) resulting from overlapping action sets when a person belongs to two or more action sets. This two-dimensional space may be more or less "full" or "empty" (see Figure 6.3).

Outer Zone: The space created by the set of persons in an individual's action set who are not directly linked with persons in the network's inner zone (see Figure 6.4).

Bridge: A person in the network's outer zone who belongs to two inner zone action sets and thus may indirectly link two individuals in the inner zone not themselves directly linked.

Network: The structure of the set of action sets created by direct or indirect links between persons in its inner and outer zones (see Figure 6.5).

Knit: The texture of the network's structure which may be more or less "tight" or "loose." It is measured by indicators called density, centrality, reciprocity, and hierarchy.

NOTES

1. This section of the chapter is a condensation of a paper, "Circles Around the Circle: A Preliminary Report on a Network Approach to Representation at the Local Level," prepared for presentation at the annual meeting of the Midwest Political Science Association in Chicago, April 12-14, 1984. The paper was coauthored by Lawrence Rothenberg and acknowledges its indebtedness to James Kuklinski, my collaborator on the Redwood Network Project. The research is funded by the National Science Foundation under Grants SES-80-17766 and SES-83-09859.

2. The networking question was open-ended, and the respondent could or could not volunteer a response that tied him or her directly to a council member. We did this because we are sensitive to the fact that a closed question easily creates an artifact: If you ask citizen Q to give his or her perceptions or impressions of a representative, he or she is likely to be obliging and give some response. But what this response means in the real-world representational relationship is difficult to say. Nevertheless, as pointed out in the text, we also asked a "name recognition" question and another about the respondent's personal acquaintance with recognized council members.

3. The project's "federal design" is a major aspect of the entire research we are conducting. It is based on theoretical notions articulated by Eulau (1973) to the effect that a full understanding of representation in the United States really calls for analysis at the three major levels of government—local, state, and national. We had to give up the state level because the resulting length and repetitiveness of the interview made it prohibitive.

4. Anthropologists fuss a great deal about network concepts. We cannot possibly discuss all the arguments here. For a justification of the notion of "action set" see Mayer (1966).

5. Network analysis has proved useful in the study of interorganizational relations. Here also, boundaries are relatively easy to define in advance of research (see Galaskiewicz, 1979, and the literature cited there).

6. Contemporary network analysis works with essentially three types of models—interaction, structural equivalence, and spatial. We have been influenced by all three types but adapted their formulations to the real-world conditions of survey analysis in the general population (for a short and precise overview, see Knoke and Kuklinski, 1982).

7. The question: "We hear a lot of talk these days about liberals and conservatives. Would you please look at the seven-point scale here on which the views that people might

hold are arranged from extremely liberal to extremely conservative? When it comes to *social issues*, where would you place yourself on this scale, or haven't you thought much about it?" The same question was asked concerning "economic issues."

8. The questions were asked in connection with the relevant network questions and read, "Would you favor or oppose bringing this industry into your community?" and, "In general, would you favor or oppose removing these books from the library?" The mean scores for the networks ranged from 1 (favor) to 5 (oppose).

9. The coefficient of variability is particularly useful for measuring a network's relative homogeneity or heterogeneity in attitudes and perceptions because it reflects similarities and differences within the network. As the networks may have very different means, it would be "somewhat misleading to compare the absolute magnitudes of the standard deviations" as measures of homogeneity. Blalock (1960: 73) notes: "One might expect that with a very large mean he would find at least a fairly large standard deviation. He might therefore be primarily interested in the size of the standard deviation *relative to that of the mean*. This suggests that we can obtain a measure of the relative variability by dividing the standard deviation by the mean." The coefficient of variability $V = \frac{s}{\bar{x}}$. The lower V is, the more homogeneous or less heterogeneous is the network as a whole on whatever property is being measured.

REFERENCES

ALMOND, G. A. and S. VERBA (1963) The Civic Culture: Political Attitudes and Democracy in Five Nations. Princeton: Princeton University Press.

BELOFF, M. [ed.] (1948) The Federalist or, The New Constitution. Oxford: Blackwell.

BLACK, G. S. (1970) "A theory of professionalization in politics." American Political Science Review 64: 865-878.

———(1972) "A theory of political ambition: career choices and the role of structural incentives." American Political Science Review 66: 144-159.

BLALOCK, H. M. (1960) Social Statistics. New York: McGraw-Hill.

COLEMAN, J. S., E. KATZ, and H. MENZEL (1966) Medical Innovation. Indianapolis: Bobbs-Merrill.

CRENSON, M. A. (1978) "Social networks and political processes in urban neighborhoods." American Journal of Political Science 22: 578-594.

CRONIN, T. E. (1970) "Metropolity models and city hall." Journal of the American Institute of Planners 36: 189-197.

DIESING, P. (1972) Patterns of Discovery in the Social Sciences. London: Routledge & Kegan Paul.

DYE, T. R. (1966) Politics, Economics, and the Public. Chicago: Rand McNally.

EULAU, H. (1971) "The legislative system and after: on closing the micro-macro gap," in O. Walter (ed.) Political Scientists at Work. Belmont, CA: Duxbury Press.

———(1973) "Polarity in representational federalism: a neglected theme of political theory." Publius—The Journal of Federalism 3: 153-195.

———(1977) "Multilevel methods in comparative politics." American Behavioral Scientist 21: 39-62.

———(1980) "The Columbia studies of personal influence: social network analysis." Social Science History 2: 207-228.

———(1984) "The Redwood Network Project: small-scale research at the local level." ICPSR Bulletin 4, 2: 1-2.

———and R. EYESTONE (1968) "Policy maps of city councils and policy outcomes: a developmental analysis." American Political Science Review 62: 124-143.

———and K. PREWITT (1973) Labyrinths of Democracy: Adaptations, Linkages, Representation, and Policies in Urban Politics. Indianapolis: Bobbs-Merrill.

———and L. ROTHENBERG (1986) "Life space and social networks as political contexts," in H. Eulau, Politics, Self and Society. Cambridge, MA: Harvard University Press.

———and J. W. SIEGEL (1981) "Social network analysis and political behavior." Western Political Quarterly 34: 499-509.

———J. C. WAHLKE, W. BUCHANAN, and L. C. FERGUSON (1959) "The role of the representative: some empirical observations on the theory of Edmund Burke." American Political Science Review 53: 742-756.

EYESTONE, R. (1971) The Threads of Public Policy: A Study of Policy Leadership. Indianapolis: Bobbs-Merrill.

———and H. EULAU (1968) "City councils and policy outcomes: developmental profiles," in J. Q. Wilson (ed.) City Politics and Public Policy. New York: John Wiley.

FENNO, R. F., Jr. (1978) Home Style. Boston: Little, Brown.

GALASKIEWICZ, J. (1979) Exchange Networks and Community Politics. Beverly Hills, CA: Sage.

GOODMAN, L. A. (1961) "Snowball sampling." Annals of Mathematical Statistics 32: 148-170.

GRANOVETTER, M. S. (1973) "The strength of weak ties." American Journal of Sociology 78: 1360-1380.

———(1976) "Network sampling: some first steps." American Journal of Sociology 81: 1287-1303.

HERSON, L.J.R. (1957) "The lost world of municipal government." American Political Science Review 51: 330-345.

HOFFERBERT, R. I. (1966) "The relationship between public policy and some structural and environmental variables in the American states." American Political Science Review 60: 73-82.

HUCKFELDT, R. R. (1983) "Social contexts, social networks, and urban neighborhoods: environmental constraints on friendship choice." American Journal of Sociology 88: 651-669.

HUNTER, F. (1953) Community Power Structure: A Study of Decision Makers. Chapel Hill: University of North Carolina Press.

JANOWITZ, M. (1952) The Community Press in an Urban Setting. New York: Free Press.

KNOKE, D. and J. H. KUKLINSKI (1982) Network Analysis. Beverly Hills, CA: Sage.

LAUMANN, E. O. (1973) The Bonds of Pluralism. New York: John Wiley.

———and J. P. HEINZ (1984) Chicago Lawyers: The Social Structure of the Bar. New York: Russell Sage Foundation.

———and F. U. PAPPI (1976) Networks of Collective Action. New York: Academic Press.

LOVERIDGE, R. O. (1971) City Managers in Legislative Politics. Indianapolis: Bobbs-Merrill.

LUPSHA, P. A. (1967) Swingers, Isolates and Coalitions: Interpersonal Relations in Small Political Decision-Making Groups. Ph.D. dissertation, Stanford University.

MAYER, A. C. (1966) "The significance of quasi-groups in the study of complex societies," in M. Banton (ed.) The Social Anthropology of Complex Societies. London: Tavistock.

McCALLISTER, L. and C. S. FISCHER (1982) "A procedure for surveying personal networks," in R. S. Burt and M. J. Minor (eds.) Applied Network Analysis. Beverly Hills, CA: Sage.

MILLER, W. E. and D. E. STOKES (1963) "Constituency influence in congress." American Political Science Review 57: 45-56.

OLSON, M., Jr. (1965) The Logic of Collective Action. Cambridge, MA: Harvard University Press.

PITKIN, H. F. (1967) The Concept of Representation. Berkeley: University of California Press.

PREWITT, K. (1970) The Recruitment of Political Leaders: A Study of Citizen-Politicians. Indianapolis: Bobbs-Merrill.

———and H. EULAU (1969) "Political matrix and political representation: prolegomenon to a new departure from an old problem." American Political Science Review 63: 427-441.

SALISBURY, R. H. (1968) "The analysis of public policy: a search for theories and roles," in A. Ranney (ed.) Political Science and Public Policy. Chicago: Markham.

SCHEUCH, E. K. (1969) "Social context and individual behavior," in M. Dogan and S. Rokkan (eds.) Quantitative Ecological Analysis in the Social Sciences. Cambridge, MA: MIT Press.

SHARKANSKY, I. (1969) The Politics of Taxing and Spending. Indianapolis: Bobbs-Merill.

SHEINGOLD, C. A. (1973) "Social networks and voting: the resurrection of a research agenda." American Sociological Review 39: 712-720.

STONE, C. N. (1980) "Systemic power in community decision making: a restatement of stratification theory." American Political Science Review 74: 978-990.

TRUMAN, D. B. (1951) The Governmental Process. New York: Knopf.

WAHLKE, J. C., H. EULAU, W. BUCHANAN, and L. C. FERGUSON (1963) The Legislative System: Explorations in Legislative Behavior. New York: John Wiley.

WEISSBERG, R. (1978) "Collective vs. dyadic representation in congress." American Political Science Review 72: 535-547.

ZISK, B. H. (1973) Local Interest Politics: A One-Way Street. Indianapolis: Bobbs-Merrill.

7

Rethinking Who Governs?

New Haven, Revisited

ROBERT A. DAHL

Time appears to have borne out the prediction advanced by a contributor to this volume in an early review of Dahl's Who Governs?: Democracy and Power in an American City *(1961). In a review of the Woodrow Wilson Award winning book for 1962, Eulau wrote in the* American Political Science Review: *"I have no doubt that the book will join the classical community studies as a classic in its own right" (1962: 144). Indeed, it has. Along with Hunter's study of Atlanta (1953), Dahl's* Who Governs? *is—25 years after its publication—still the centerpiece of American community power studies. The present chapter presents a brief explanation of the New Haven study and an interview with Dahl in which he analyzes the methodology of* Who Governs?, *responds to critics of the New Haven study, and suggests some new directions for community power research and for the study of local government generally.*

METHODOLOGY

There are five major methodological "ingredients" to the *Who Governs?* study. The first of these is issue identification. Dahl and his associates identified education, redevelopment, party nominations, and charter reform as major issues to study in New

Haven. Second, Dahl attempted to identify the major decisional actors. This involved an attempt to identify actors who initiated policy, as well as to establish a "box score" of wins or losses for the various policy actors. As the following interview makes clear, Dahl first attempted to identify such actors via the reputational method of Hunter. Third, Dahl identified the social and economic notables of New Haven using a variety of means, including listing the 50 top property owners (using assessed valuation as a measure), society page lists of those attending Assembly and cotillion functions, city directories, company reports, *Poor's Register of Directors and Executives*, and *The Director's Register of Connecticut* (1958). Fourth, Dahl directed four surveys, including (1) a questionnaire originally sent to 1063 policy actors ("leaders" and "subleaders"); (2) a follow-up random sample of 286 leaders and subleaders; (3) a November-December survey of 197 randomly drawn registered New Haven voters following the defeat of a proposed new city charter; and (4) a final interview of 525 voters drawn from a list of registered voters, in which—unlike the earlier voter survey—nonvoters were interviewed instead of being replaced by randomly drawn voters in a voter-only survey. The fourth survey was supervised by William Flanigan and conducted in the summer of 1959. A fifth and final check on the *Who Governs?* study was a massive data-gathering project in which the group attempted to collect returns for all known New Haven elections through official records and newspaper accounts, as well as historical studies. This breakdown of the vote by party, candidate, office, and ward was correlated with socioeconomic characteristics for each ward, including the numbers of foreign-born by country of birth, the number of white and black residents, the number of males, and (since 1930) females of voting age.

FINDINGS

Briefly, *Who Governs?* had eight major findings. First, Dahl found little overlap between social, economic, and decision "notables" in New Haven. Instead, influence in New Haven was extremely specialized and limited to one or a few areas of influence and access. Second, the black population, while clearly denied opportunities in many aspects of the social and economic life of New Haven, were overrepresented in New Haven electoral politics relative to their numbers in the overall voting population. Third, Dahl stressed the importance of elected officials, especially the mayor, in shaping issues such as adopting "big" versus "little" redevelopment strategies. Fourth, and reminiscent of the first finding, Dahl found that only a small number of people had direct influence (either to initiate or to veto proposals) but many, including voters, had an indirect influence on community decisions. Fifth, as Dahl discusses in the following interview, leaders in New Haven worked within the context of a "democratic creed" and a set of latent predispositions and community collective experiences that bestow advantages and disadvantages to players, as well as help to shape outcomes, procedures, and options. Sixth, Dahl did not describe New Haven as without a need for change. Rather, he labeled New Haven a "pluralistic democracy" and "a republic of unequals." Dahl elaborates on these terms in the following discussion. Seventh, the most prominent theoretical contribution of *Who Governs?* was the division of New Havenites into *homo civicus* and the more politically active *homo politicus.* Finally, Dahl took a stab at typologizing and describing the patterns of coalition formation and policy leadership within local communities. His discussion of the three dominant patterns—spheres of influence, the executive-centered coalition, and rival sovereignties—remains a classic of the attempts to describe and understand the paths of influence and policymaking in local communities.

INTERVIEW: ROBERT A. DAHL

WASTE: In a set of interviews for an earlier research project (Waste, 1986), you noted that you had been impressed by Hunter's study of Atlanta and the reputational method, and had tried to replicate the reputational method in the New Haven study. I wonder if you care to elaborate on that.

DAHL: My memory of that, I must confess, is now a little hazy, as it is on many of the aspects of the New Haven study. What I do remember is that we found the attempt to use reputation simply too undiscriminating and too insensitive; too vague and general. It did seem to suggest to us that the reputations that people had bear only slight relationship to what they actually did in political life and in decisions. We did abandon that approach early on as a result of this dissatisfaction with the methodology.

WASTE: In a symposium at Notre Dame later edited and published as *Power in America* (D'Antonio and Ehrlich, 1961: 105-107), you noted in response to a question that—in your wording—any organized and legitimate interest group could have an impact on local community politics. That's come to be something of a celebrated passage. I wonder if you could elaborate on what you meant.

DAHL: I think the wording is important. A rather similar set of words also occur in the last chapter of *A Preface to Democratic Theory* (1956). By "legitimate" I meant that they were widely accepted as having a right to participate in local politics. One reason for using a criterion like this is that if a group can be defined as outside the sphere of American political life, clearly its impact is very limited, if any, or it is essentially excluded from the political process. I have in mind, for example, any group that can be effectively labeled as communist, or even "extremist." Such a group may simply be excluded from the normal political process that is open to others. Thus, a group has to be seen as legitimate in some sense to gain entry into the political system. In addition to that, in order to gain entry it has to be active—engaged in some significant way. Any active and legitimate group is likely to be heard. In *A Preface to Democratic Theory*, I said that by "heard" I meant more than simply that their voices would make a noise—more than simply heard in a technical sense. I meant that other people would begin to listen to them. Now that's not the same thing as claiming that they will necessarily prevail, that their views will win out or, certainly, that their voices will be heard as powerfully as the voices of other groups.

An important characteristic of a political system is that it be relatively permeable, that it remain open to groups who are active, organized, and want to be heard. I think that is a characteristic of New Haven and most cities in the United States, and I believe that it is substantially true of the national political system. However, it is always important to remember that when I say something like this, I generally have in mind a background of democratic theory and practice; that is to say, I tend to think of American political life in a comparative context. When you begin to think comparatively about different kinds of regimes, including nondemocratic regimes, then you will tend to think it very important to distinguish between political regimes in which almost every group has some opportunity to be heard and regimes in which lots of people—possibly most people—don't have the slightest opportunity to be heard, except, perhaps, by dynamiting buildings and officials. I lived for several months in Chile in 1970, during the good old decent days of the Chilean political system, and was deeply shocked at what happened to it in 1973 when the military took over. Just a few months ago, I returned and spent a week there—under the Pinochet regime, of course. In a very brief period of time, during 1973-1974, the Chilean regime was transformed from one in which most groups—including the Communist Party and Communist-led trade unions—were "heard," to one in which vast segments of the population could be, and were, ignored. I believe the difference in regimes is important and worth stressing, if one is thinking in broadly comparative and historical terms.

WASTE: On a related theme, you noted in the first appendix in *Dilemmas of Democratic Theory* (1982: 207) that there seems to be one view of pluralists, held primarily by critics of pluralism, that in a pluralist political system not only are there several groups—making it a "plural" system—but that the groups were more or less equally weighted, and that has come to be a view of pluralism "writ large." I wonder if you care to comment or elaborate on such a view of pluralism?

DAHL: I've always been puzzled by how that interpretation came about, or what concrete words or expressions it is based on, because I can't find a passage in my own writings anywhere that would reasonably seem to lead to the interpretation that all groups are in some sense equal. That interpretation misinterprets my views in two ways, one primarily empirical, the other primarily normative. As to the empirical mistake, it is perfectly obvious that even those that are active and legitimate are clearly not equal in influence, by almost any reasonable measure of influence. In fact, they are usually decidedly unequal. Inequality of influence—as I say in *Modern Political Analysis* (1963), a book intended to be no more than introductory—is a characteristic of virtually all political systems, in all times and all places. It's a virtually universal characteristic of political life that individuals and groups are unequal in influence. I would have thought that this empirical proposition is so axiomatic, so obvious, so utterly taken for granted that an attempt to interpret pluralist theory or any other theory as denying it would in effect be saying by implication that authors of pluralist theory, or any such theory, are utterly ignorant of the nature of the world. Because it seems to me perfectly clear that those of us who came to be called pluralists, and certainly I, have never believed that all groups in politics are equal. I have always found that view of pluralism quite puzzling. As to the normative mistake, a point that I made in the *Dilemmas* book, and before—in fact, Lindblom and I make it very explicitly in the 1953 book, *Politics, Economics and Welfare*—is that from a democratic point of view it would not necessarily be a good thing if all groups were equal. A group consisting of ten people and a group consisting of 10 million people are both groups, but it would certainly be a great violation of our conception of political equality if they had equal influences in political life. So not only is group equality patently false as a factual empirical statement, but it is patently undesirable as a normative objective. This is why I am so mystified that anyone familiar with the assumptions that virtually any political scientist is prepared to make, and certainly any political theorist, would suppose that pluralists have contended that in New Haven or anywhere else in the world, all groups are roughly equal in influence.

I can think of three possible sources for that mistaken interpretation. A careless reading of what I have written, or reading it out of context—although it would take a terribly careless reading, even so—might lead a reader to conclude that I was asserting something like equality of influence. Several persons interpreted a passage in *A Preface to Democratic Theory* (1956) along the lines we are discussing. I have to conclude that they failed to read everything around the passage in order to misunderstand me so badly—which I find disappointing, since I write very carefully and I like to think that people will read with equal care. A second source for this misinterpretation may be that you can read Truman's *The Government Process* (1951), which is, of course, on group theory, and can read it as implying some such notion. If you read the book carefully, however, such a view is explicitly contradicted by what Truman has to say about the inequality of different groups.

WASTE: Truman does, for example, have a whole passage (1951: 14-17, 33-44, 506-519) on "strategic access" as being unequal for groups.

DAHL: Exactly. You would have to read the book and ignore whole passages such as that one. Although it's a difficult book to read, if you read it carefully you would never come up with such interpretations. A third source is a more theoretical difficulty. It has appeared to me, and again I find this puzzling, that people often find it hard to think in terms of what might be called degrees of inequality. There is a tendency to think that if you depart from equality at all, you must have domination. And if you don't have domination, then you must have equality. Now, we all know that that's simply not true. We can have inequality, greater inequality, and lesser inequality. And, it seems to me, much of *Who Governs?* and much of what I've written elsewhere is an attempt to provide qualitatively, provide qualitative descriptions (since we lack the quantitative indicators that would serve the purpose) of political systems which cannot be accurately described either as systems of domination or systems of equality. It is a very complex matter to describe regimes that are somewhere between these extremes, particularly since different systems may lie at different points along the hypothetical continuum between equality and domination.

WASTE: Speaking of *Who Governs?*—you note in *Who Governs?* and elsewhere that there is, in your words, a "democratic creed" that limits leaders and helps shape consensus. Do you think that is still an active force, and is still as strong a force in New Haven and local communities generally as when you were writing in the 1960s?

DAHL: Well, again, let me answer in two parts. The first, directed more generally to reasons why I stressed the importance of that notion and, second, to the question about New Haven in particular. My reason for stressing the importance of the democratic creed is that political culture is a terribly important background factor in the nature of political regimes. Anybody who tries to come to grips with the question of why democracy—or I would say polyarchy—exists in some countries and not in others sooner or later is likely to conclude that the belief system—the habits, the mores, the customs of a people, as Tocqueville said, are enormously important. Political culture, if you like, is a factor that everyone finds extremely difficult to tie down precisely, and methodologies for dealing with it are not very satisfactory. Nevertheless, if one tries to explain why democratic orders or polyarchal institutions exist in some places or at some times and not in other times and places, one is driven sooner or later to the belief system—the belief in the legitimacy of democracy, for example, or the absence of a widely shared belief in the legitimacy of democracy. In *Polyarchy* (1971) I tried to present some evidence on that matter. What I and others have called the democratic creed is a set of beliefs on democracy, relatively widely held, if unequally held and diffused, in the population of the United States. And this system of beliefs is a strategically important factor in accounting for why democratic institutions exist and have persisted and do not in many other countries. Some such set of beliefs is possibly a necessary condition, though definitely not a sufficient condition, and unquestionably an extremely important condition for democratic government.

Now, about New Haven, has that creed grown stronger or weaker in the interval since I studied New Haven? I don't have any evidence to go on to make such a judgment. I really have no way of making a judgment about possible changes in strength of the democratic creed. I wish there was an adequate baseline indicator from the time of our studies so that one could now make a study of New Haven and determine some of the things associated with the strength of the democratic creed, but that's not possible, at any rate not in a very rigorous and systematic way. I have no reason to think that it's grown any weaker, nor do I have any reason to think that it's grown significantly stronger. The demographic changes since the *Who Governs?* study are interesting but not really all that suggestive of any shift in the democratic creed as such. So I guess, Bob, the answer to your question is that I really don't know.

WASTE: Still staying on the general topic of New Haven—you described New Haven as a pluralistic democracy. What do you mean by that term? Could you elaborate on it a bit? What does it mean when you say New Haven, or any city, is a pluralist democracy?

DAHL: I used the term in *Who Governs?* maybe five or or six times, and used it without at any point explicitly indicating what it was I meant by that term. Again, in *Politics, Economics and Welfare* (1953) Lindblom and I have explicitly defined pluralism as referring to social pluralism, meaning the existence of relatively autonomous units in society fitting within the framework of a larger system. It was some such meaning that I had in mind in describing New Haven. By referring to democracy, I was using the term in a rather loose sense, meaning the presence of a set of institutions which I did not then spell out, though I had earlier, and which I would now be inclined to refer to as the institutions of polyarchy. These are fairly well defined. You certainly probably know more than anyone, including me, about how to go about defining that set of institutions. Pluralist democracy therefore joins the term democracy, meaning the institutions of polyarchy, with social pluralism, meaning the existence of groups and organizations that are relatively autonomous with respect to one another, and with respect to the state and the government.

WASTE: In fact, the last three questions tie in with each other. When speaking of a pluralist democracy, one has the conditions that you were talking about which presuppose active and legitimate interest groups, as well as the existence of a democratic creed.

DAHL: I think that's right. What I mean by pluralist democracy is, of course, much more fully developed in *Dilemmas* (1982), in which I finally spell out its meaning—although I had actually spelled out earlier what I mean by it, in one respect or another, and what problems are inherent in that kind of regime. The clearest explanation is probably in *Dilemmas* (1982).

WASTE: The view of pluralist cities that you are presenting is a view that seems a lot more elaborate than one finds in several critics of pluralism—a view that sometimes seems reduced to a city in which lots of groups are fighting it out. Therefore, it must be a pluralist city.

DAHL: I think that's right. You can have a pluralist system without its being democratic. Juan Linz, for example, a leading authority on authoritarian regimes, shows that they can also be moderately pluralistic. I suppose Franco Spain was a case of limited pluralism in the later years. So you could have relatively pluralist systems that are authoritarian. You can also—theoretically anyway—have a monistic democratic regime where no autonomous groups play a significant role in political life. That may have been the ideal, and to some extent the practice in Greek democracy—and judging from Pocock's *The Machiavellian Moment* (1975), it was also to some extent the ideal in the republican tradition of the Middle Ages and the Renaissance. For example, the Italian republics generally did not allow factions, or at least tried not to allow factions, even though factions usually broke through the facade of unity around the common good. The same thing can be seen in Rousseau's famous hostility to associations in the *Social Contract* which reflects the older view. In principle, then, you could have a democratic system that wasn't pluralistic, and an authoritarian system that was pluralistic without being democratic. However, I think that it would be impossible for a modern state, at least at the national level, to have polyarchal institutions and yet not be pluralistic.

WASTE: Still on the *Who Governs?* study—what aspects of the book have you found most and least satisfying over time?

DAHL: I think that it's a very well-crafted book, and the craftsmanship of the book has held up. Of all the things that I have written, it may be one of the most nicely written and put together. That's one reason, I think, why people come back and read the book many years later—because they enjoy reading it. I'm pleased with that quality.

I think the general observations about New Haven as it then was—and this is certainly subject to an enormous amount of dispute—have held up pretty well over time. New Haven was and is a community in which a diversity of groups exist and bear on the making of public policy. I must say that I'm not the best judge of my own work, and somebody else could answer that question better than I could.

WASTE: What about the research design? It seems to me that there's an argument to be made that the research design has also held up very well over time. Considering that you conducted surveys of the subleaders and of registered voters, and that you focused on the most controversial and/or—in some cases—the most expensive decisions facing the city as a combined way to capture what's going on in a city, do you think that you would change the methodology if you were conducting the study over today? Or would you retain the methodology as it is?

DAHL: I might change it a bit, but, like you, I think that the approach in *Who Governs?* was a rich array of techniques, a very diverse methodology. One of my disappointments was that my professional colleagues often have not fully grasped the diversity of techniques that we used. We were probing at that beast with lots of different instruments—including, I should mention here, the instrument named Roy Wolfinger who was sitting in the mayor's office. He was a source of very important information. Each of those techniques had its place and was important in the interpretation of the book. Again, for some reason, a number of readers seem to have come to the conclusion that all we did was study three decisions. That is very far from the case.

I think that it was a good methodology. However, if I were going to redo it, there are some things that I might now do differently. I have a sense that it would be much more difficult and expensive to do the study today. We did it on a shoestring at the time. Now it would be enormously more difficult methodologically because one would be doing them on top of all the methodological critiques that followed the publication of *Who Governs?* There might be many more questions on the surveys, and the survey itself would be a much more expensive project to do today.

WASTE: Admittedly, it would probably be a lot more expensive replicating the earlier study today. Still, I don't know how many more ways you could poke into the decisional arena of a city other than: (1) survey research, (2) observers and interns, (3) following the expenditures of large sums, and (4) asking the people you can definitely identify as being participants in what happened. I can't really think of a fifth or sixth way that you could use to triangulate what's going on in cities.

DAHL: One of the methodologies that you didn't mention and which, again, I took some pleasure in, is the attempt to put this in a historical setting, and to tease out changes from the historical record. Of course, this presented some problems in determining what was reliable, but we attempted to learn about changes over time from the historical record. That is an important aspect which community studies ought to follow where they can. New Haven, of course, had lots of historical records and accounts. Many communities would not have good historical records.

WASTE: I started to say that you could have asked nonvoters, but in fact you did that on the second survey to the extent to which you asked registered voters who had not voted in that election to answer the question, and you didn't throw those responses out as you did with the earlier survey.

DAHL: I think that there are some subjects that can be tackled that I would now be tempted to take on. Again, possibly, at the price of making it a bigger, more sprawling, and different kind of project.

WASTE: On the dimension of studying community power or local government—generally, what do you regard as the most promising approaches to studying community power in the near future, and why?

DAHL: Two things ought to be done, although they are enormously difficult to do. One is something that you yourself were mentioning a moment ago and in which I heartily concur. We badly need more systematic comparative studies. Even if not systematic studies, studies which make comparison possible. Or separate studies that make comparison possible. One of the grave defects of New Haven is that it's a case study. It's a single case. There's no getting around it, no possible way anyone can deny that. It has all the riches and advantages of a single case study, it has all the limitations—even if we are prepared to assume, as of course I am, that the characteristics described in *Who Governs?* hold for New Haven, the attempts to generalize from New Haven are pretty risky. We need to know the respects in which other communities vary from New Haven, and the causal or explanatory factors that would account for these differences. That is something that you have been more deeply engaged in than anybody I know. I have come more and more to the conclusion that, indeed, all study of government needs to be comparative. The study of the American political system really should be comparative and not simply a study of American politics. Without a comparative focus, you can't really understand how the American system of politics is similar and how it differs from other regimes, nor can you explain the similarities and differences.

Now there is a second way in which I might change the study if I were to do it over today. I do think, and here my critics have a good point, that while there is probably more in *Who Governs?* than some critics have paid attention to about the limiting factors on the decision makers, it would seem useful to say more about the limits. Some of them I took for granted, but I am not sure it's a good idea to take them for granted; perhaps it would be better to spell out the limitations. For example, I took for granted that New Haven, like any other city in the United States, operates as a part of a capitalist system—an economic system in which many crucial decisions are made with some degree of autonomy, though not necessarily local autonomy, by corporate leaders. That's simply the nature of a capitalist system—or even more broadly, it's the nature of any market system, whether capitalist or socialist. That being the case, not only are some decisions important in the life of the community that lies outside the conventionally defined political realm of municipal politics, but also within the realm of the conventionally defined political life of a municipality, decision makers will be influenced—directly and indirectly—by the need to take into account the decisions and influences of the market system. A lot of that I took for granted. To some extent, perhaps, *Politics, Economics and Welfare* (1953) provided me with many assumptions for decision making that I should have made explicit.

Another factor which I think that I didn't pay sufficient attention to is the extent to which New Haven is part of a larger political order. Clearly, that is extremely important—notably important during the time in which I was writing, and important at any time. For local governments, the relations with external governments are a matter of such great importance that they ought to be built into the study more fully than I did.

The third feature, which, as you have mentioned, I touched on in *Who Governs?* but which needs to be built on more extensively—even though it is enormously difficult—is the effect of the belief system. This would involve a range of beliefs that would include the democratic creed, whatever that may be, but also other kinds of beliefs about the legitimacy of capitalism, private enterprise, property, and other beliefs that limit what municipal governments and local politicians can do. Their sphere of alternative courses of action is obviously going to be limited by belief systems. More attention to ways that the belief system limits the political agenda and the alternatives open to politicians is an important thing to study. Now all that would make a pretty big study. There's a danger, especially in the third of these, that a book would turn into nothing more than a study of political culture and political belief systems. It could turn into a work on anthropology or cultural history, and a political theorist's endless quagmire.

Another point: I realize now how unusual the strengths of [Mayor] Dick Lee as a coalition-builder were at a critical juncture in the life of a city. He was highly unusual, in his personal qualities, his historical possibilities, and the outside limitations he faced. All of these variations need to be taken into account.

WASTE: At the time that you wrote *Who Governs?* political science didn't have a vocabulary to describe actors like Lee—even though Lee's type clearly wasn't unique in American history. With the discussion of figures like policy entrepreneurs, it is much easier to understand Lee's role, as you say, as not being unlimited but as being forceful and critical. The recent discussions of "policy entrepreneurs"—such as Robert Moses's role in New York City (Caro, 1975; Jones, 1983)—are a case in point. Clearly, it would have made a difference if Moses were not an entrepreneur, and it would have made a difference if some other mayor who was not an entrepreneur was in office at the time instead of Lee.

DAHL: That's why comparative studies of cities would be fruitful in enriching the typologies of actors and types of cities. Fruitful, too, just in describing them, how they come into play, why they are rare sometimes, and in some places, and so on.

WASTE: I think to follow up what you were saying about implicit things in *Who Governs?*—and maybe even in pluralist writing in general—there may be a small cottage industry in writing to make explicit what is implicit in pluralist writing. Not because, it seems to me, that it is particularly vague in any of the writings, but because it is just so fundamentally misunderstood by people who are reading them. It becomes, for example, necessary for Stone (1980) to write his article on "systemic power"—I would have thought that it was fairly clear that Mayor Lee and other city decision makers were not operating in a power vacuum in which he could do anything that he wanted.

DAHL: I thought that all those things were clear too. Any discrete political system is open at the edges—that is, there are an indefinite number of tops, bottoms, sides, an indefinite number of things going on that are connected to it. You can't describe them all. You have to take some things for granted. Nevertheless, those things that are going on around it and interpenetrate it are extremely complex. There will always be things that you will leave out that can be said to be in the systems—and to be of considerable importance. I think that's inherent and inescapable in the nature of a political system, especially a municipal system that is embedded in a larger society and a larger polity.

WASTE: Maybe this is just a short epoch in community power studies, in which people studying urban or local units have a need for more clarity and definition. Jacobs (1984) just wrote her new book on how world economies impinge on local governments—which I would have thought was obvious. Maybe there is some need to spell out the ecosystem that surrounds cities and affects actors within it.

DAHL: There probably is. I have the sense—we probably all do—that as the United States becomes increasingly a part of an international economy, as it becomes more embedded in international systems of various kinds, the need to at least identify these external factors (if for no other reason than to indicate that you are holding them to one side) will become very important, even at the national level.

WASTE: Especially, I would think, as the scale of cities becomes larger and population increases. In the United States alone, there are several cities with populations that exceed those of states entitled to two senators. Certainly New York City, for example, is larger than Montana, Alaska, or Nevada.

DAHL: Sure. Well, let's take a guess. New York City is probably larger than 40% of the countries of the world.

WASTE: And ironically, they have no representation in the upper chamber of the national legislature. If it were a country, it would have the budget of about the fifteenth largest country in the world.

DAHL: That's why at the subnational, national, and international levels, we are going to have to pay more attention to the comparative dimension of governments, to focus on their increasing interconnectedness, and to sketch this out so that we have a clearer picture of the ecosystems of local governments and how these ecosystems relate to and are influenced by the national and international systems in which they are embedded.

REFERENCES

CARO, R. (1975) The Power Broker: Robert Moses and the Fall of New York. New York: Vintage.

D'ANTONIO, W. and H. J. EHRLICH [eds.] (1961) Power and Democracy in America. South Bend, IN: Notre Dame University.

DAHL, R. A. and C. LINDBLOM (1953) Politics, Economics and Welfare. New York: Harper & Row.

———(1956) A Preface to Democratic Theory. Chicago: University of Chicago Press.

———(1961) Who Governs?: Democracy and Power in an American City. New Haven: Yale University Press.

———(1963) Modern Political Analysis. Englewood Cliffs, NJ: Prentice-Hall.

———(1971) Polyarchy: Participation and Opposition. New Haven: Yale University Press.

———(1982) Dilemmas of Pluralist Democracy: Autonomy vs. Control. New Haven: Yale University Press.

EULAU, H. (1962) "Review of Robert A. Dahl, *Who Governs*?" American Political Science Review 56: 144-145.

HUNTER, F. (1953) Community Power Structure. Garden City, NY: Doubleday.

JACOBS, J. (1984) "Cities and the wealth of nations: a new theory of economic life." Atlantic Monthly 253 (March): 41-66.

JONES, B. D. (1983) Governing Urban America: A Policy Focus. Boston: Little, Brown.

POCOCK, J. G. (1975) The Machiavellian Moment: Florentine Political Thought and the Atlantic Republican Tradition. Princeton: Princeton University Press.

STONE, C. (1980)"Systemic power in community decision making" American Political Science Review 74 (December): 978-990.

TRUMAN, D. B. (1951) The Governmental Process. New York: Knopf.

WASTE, R. J. (1986) Power and Pluralism in American Cities: Researching the Urban Laboratory. Westport, CT: Greenwood Press.

PART IV

Conclusion

8

Community Power Research

Future Directions

ROBERT J. WASTE

As the preceding pages demonstrate, we are entering a new era in community power studies. After three decades of deadlock and acrimony, the study of power in local communities has begun to show signs of both maturation and innovation—maturation in terms of the willingness of many of the major participants to enter into a serious dialogue with one another, and innovation in terms of new ways of conceptualizing the leading theoretical and methodological issues in the field. Both of these developments bode well for community power studies.

THE COMMUNITY POWER "QUESTION"

One of the few certainties in the community power debate is the importance of the question at issue. At its base, the community power inquiry is an attempt to understand and describe the distribution and use of political power in democratic societies. While the main research laboratories of community power researchers are primarily local governmental units in the United States, the community power question as such is really an attempt to empirically describe and to normatively understand and assess the quality of life in a democratic society. Reduced to its bare essentials, the questions in each case are: "Is the polity in community *x* a democratic polity? How strongly democratic (or undemocratic) is it? Further, does this assessment hold true across issues and over time? Finally, how does this polity compare with other researched polities? In other words, how does this polity rank along a continuum of known democratic and antidemocratic regimes?" Viewed in this light, community power research is inseparable from the larger realms of both democratic theory and comparative government. Hence, an advance in one of these fields is necessarily an advance in the others.

RECENT ADVANCES IN COMMUNITY POWER RESEARCH

As the preceding chapters illustrate, community power research is moving forward on two fronts. First, as the Cold War thaws, it may become increasingly possible for community power researchers of both the pluralist and elitist persuasions to talk with—and not at—each other. As with all such Cold War efforts at détente, the duration and benefits (both short and long term) are difficult, if not impossible, to assess—particularly in the early stages of the would-be rapprochement. Certainly, the central question at issue—whether New Haven or other U.S. cities are elitist or democratic in character—will not disappear. Nor should it, since the question itself provides much of the raison d'être for community power research. Rather, the probable changes—both in the short and long term—would appear to be the opportunity for both students and professional researchers in the field to assess the recent changes in both theory and methodology, and to test these advances to determine if they offer advantages to those seeking to empirically describe and to normatively assess the character and quality of life in American cities.

A second front or avenue of change in community power studies is the increasing number of recent innovations—both theoretical and methodological—in the community power field. The present volume contains much of what is most valuable in these innovations. As we noted in Chapter 1, the contributors in this volume are building on a base of recent and substantial innovation in the larger community power field itself. Notable innovators in this sense include Floyd Hunter (1980), Nelson Polsby (1980), Gunnar Falkemark (1982), Peter Trounstine and Terry Christensen (1982), John Manley (1983), Mark Mizruchi (1984), Harvey Molotch (1976, 1979, 1984), and Fred Wirt (1984).

ASSESSING THE CHANGES

The changes in community power conceptualization and methodology represented in the present volume are considerable. Taken as a whole, they are characterized by a drive for linkage, clarity, and methodological precision. The chapters by Dye and Dahl both feature linkage as a central issue. Dye demonstrates successfully the benefits of linking community power research to the larger concerns of public policy studies, while Dahl presents a forceful case for placing community power research in a more comparative overall context. Linkages such as these are not only beneficial in the abstract intellectual sense but might also help to guard against the narrow parochialism that has characterized much of earlier community power research.

The chapters by Stone and Waste represent an attempt to clarify and operationalize basic concepts in community power research. Parenthetically, it should be added, these chapters also reiterate the linkage theme stressed in the earlier chapters. It is difficult to see how an analysis of such concepts as "power" and "pluralism" would not be of relevance—not simply to the community power researcher but also to the larger social science community.

Finally, the chapters by Domhoff and Eulau demonstrate both the increasing sophistication of community power research and the increased need for still more methodological sophistication. Clearly, both methodologies are positive movements forward. And, as with all methodologies, both require comparative data and further refinement to be of maximum benefit to researchers in the field. The latter consideration leads us to our final point—

namely, a discussion of a future agenda for community power research.

COMMUNITY POWER: AN AGENDA FOR FUTURE RESEARCH

As noted earlier, the growth machine theory and the use of network analysis by Eulau demonstrate both the increasing refinement in community power methodologies and the need for further refinement. Specifically, the growth machine theory—and this is a problem inherent in both pluralist and elitist research—needs further empirical operationalization to be useful for comparative research. Theories per se do not prove anything. Rather, they require an explicit field methodology to allow replication in several cities and locales, and to facilitate the comparisons that are the grist of the comparative local government endeavor. In the short term, the problem for elitist researchers would seem to be to aid in the transition from an intuitively compelling theory to an empirically verifiable hypothesis. The theory requires empirical operationalization so that we may know when city *x*, for example, has a growth machine. Further, such growth machines will only be well understood in the context of comparative research—comparing the growth machine in Dallas, for example, with that in San Diego, and so on.

The problem for researchers employing network analysis is precisely the opposite of the problem with the growth machine theory. Network analysis is, on its face, a well-developed and still-developing community power research methodology. The absence here is that of an overall theoretical framework in which to place the eventual findings of multicity or macro studies in network analysis. Clearly, this is a suggested agenda for further research. The current micro project—Eulau's Redwood Network Project—shows every sign of leading to a macro or comparative study which could yield the aggregate data necessary for theory construction.

Linkage should continue to be an important item in the community power research agenda. Specifically, it is reasonable to expect that the linkage found in the chapters by Dye and Dahl would be expanded greatly in the future. The linking of community power concerns to the larger public policy, urban policy, sociological inquiry, and comparative government fields would seem extremely promising. To name a few areas of possible inquiry, links might be sought to studies in policy implementation, policy entrepreneurs, policy agenda-building, urban service delivery, municipal coproduction, and the cutback management associated with local governments undergoing fiscal stress. Alternatively, community power research might benefit from more efforts in a comparative direction—both conducting more multicity or macro studies to facilitate comparison and theory construction (and testing), and encouraging a more careful reading of applicable comparative research in democratic societies.

Concepts such as "consociationalism" (Lijpart, 1977) and "segmented pluralism" (Dahl, 1966, 1973)—originally advanced to explain European governmental scenarios—may prove useful in describing and analyzing coalition formation in local governmental units in the United States. As Dahl notes in his interview, several U.S. cities exceed the size of modern nation-states. Given this blunt similarity in terms of size and scale, comparative research would appear to be a fruitful hunting ground for community power research "links." However, size is not the controlling variable in seeking comparisons of different political cultures. Difficulties exist in comparing any local political units. Researchers might profitably draw on the Sunbelt/Frostbelt studies (Abbott, 1981; Sawers and Tabb, 1984) to determine if any differences in networks, coalition patterns, growth machines, or policy preferences or styles exist that are unique to each region, and that lead to substantive differences in community politics and decision making. I would argue, for example, that Frostbelt cities are more likely to be pluralist in character than faster-growing Sunbelt cities. As Abbott (and Sawers and Tabb) has argued, neighborhoods and downtown progrowth business groups have simply been at it longer in Frostbelt cities; they have had more time to practice coalition formation and to benefit from past attempts to influence local policymakers.

CONCLUSION

In sum, community power research is on the verge of a new era. As projected in the present volume, this new era encompasses a willingness by pluralists and elitists to work together to develop: (1) a more useful and precise vocabulary for community power research, (2) more exact and replicable field methodologies to describe and analyze power in local communities, and (3) more explicit and self-conscious linkages between the study of community power and related fields of social science generally. It remains to be seen, of course, how useful in the long run such a new era in community power may be to understanding the relations of leaders and citizens in democratic polities.

REFERENCES

ABBOTT, C. (1981) The New Urban America: Growth and Politics in Sunbelt Cities. Chapel Hill: University of North Carolina Press.

DAHL, R. A. [ed.] (1966) Political Oppositions in Western Democracies. New Haven: Yale University Press.

———[ed.] (1973) Regimes and Oppositions. New Haven: Yale University Press.

FALKEMARK, G. (1982) Power, Theory and Value. Lund, Sweden: Liber Gleerup.

HUNTER, F. (1980) Community Power Succession: Atlanta's Policy Makers Revisited. Chapel Hill: University of North Carolina Press.

LIJPART, A. (1981) Democracy in Plural Societies. New Haven: Yale University Press.

MANLEY, J. (1983) "Neo-pluralism: a class analysis of pluralism I and pluralism II." American Political Science Review 77 (June): 368-383.

MIZRUCHI, M. S. (1984) "An interorganizational model of class cohesion." Power and Elites 1 (Fall): 23-36.

MOLOTCH, H. (1976) "The city as a growth machine." American Journal of Sociology 82 (September): 309-330.

———(1979) "Capital and neighborhood in the United States." Urban Affairs Quarterly 14 (March): 289-312.

———(1984) "Romantic Marxism: love is still not enough." Contemporary Sociology 13 (March): 141-143.

POLSBY, N. W. (1980) Community Power and Political Theory (2nd ed.). New Haven: Yale University Press.
SAWERS, L. and W. TABB (1984) Sunbelt Snowbelt: Urban Development and Regional Restructuring. Austin: University of Texas Press.
TROUNSTINE, P. J. and T. CHRISTENSEN (1982) Movers and Shakers: A Study of Community Power. New York: St. Martin's Press.
WIRT, F. M. (1984) "Rethinking community power." Power and Elites 1 (Fall): 89-98.

About the Contributors

Robert A. Dahl is professor of political science at Yale University. His publications include *A Preface to Democratic Theory; Who Governs? Democracy and Power in an American City; Modern Political Analysis;* and *Dilemmas of Pluralist Democracy: Autonomy and Control.*

G. William Domhoff is professor of sociology at the University of California at Santa Cruz. His publications include *Who Rules America?; The Higher Circles; Who Really Rules?;* and *Who Rules America Now? A View for the 80's.*

Thomas R. Dye is professor of policy sciences at Florida State University. His publications include *Understanding Public Policy; Who's Running America? The Carter Years; Who's Running America? The Reagan Years;* and *Politics in States and Communities.*

Heinz Eulau is professor of political science at Stanford University. His publications include (with K. Prewitt) *Labyrinths of Democracy: Adaptations, Linkages, Representation, and Policies in Urban Politics;* and *Politics, Self and Society.*

Clarence N. Stone is professor of political science at the University of Maryland. His publications include *Economic Growth and Discontents;* and *Urban Policy and Politics in a Bureaucratic Age.*

Robert J. Waste is assistant professor and director of the Institute of Public and Urban Affairs at San Diego State University. His publications include *Power and Pluralism in American Cities* and (with D.R. Marshall) *Large City Responses to the Community Development Act: Researching the Urban Laboratory* (forthcoming).